CHEMISTRY RESEARCH AND APPLICATIONS

TERPENES

BIOSYNTHESIS, APPLICATIONS AND RESEARCH

Chemistry Research and Applications

Additional books in this series can be found on Nova's website under the Series tab.

Additional e-books in this series can be found on Nova's website under the eBooks tab.

CHEMISTRY RESEARCH AND APPLICATIONS

TERPENES

BIOSYNTHESIS, APPLICATIONS AND RESEARCH

ALBIN BJARKE
EDITOR

nova
science publishers
New York

Library of Congress Cataloging-in-Publication Data

ISBN: 978-1-53613-622-7

Published by Nova Science Publishers, Inc. † New York

CONTENTS

PREFACE

In this collection, terpenes in oregano, mainly represented by the essential oils, are extensively documented for their antimicrobial, antifungal, antiviral, antioxidant, anti-inflammatory and anticancer properties, with the goal of describing the biological properties of terpenes from oregano and their mechanism of action. Next, the authors focus on solid acid catalysts as potential heterogeneous catalysts for terpenes conversion into fragrances and flavours, replacing homogenous ones. Different reactions, such hydration, alkoxylation, esterification, isomerization and cyclization of terpenes over heterogenous catalyts will be reviewed. Afterwards, turpentine (consisting of the volatile terpenes fraction from conifer wood) is analyzed because of its place as the most widely produced essential oil. Lastly, the authors discusses the effects of terpene-rich essential oils. Although several studies describe the cardioprotective effects of these essential oils, a systematic review is lacking. Therefore, a compilation of these studies is presented with special focus on the hypotensive, vasorelaxant, modulation of ion channels, anti-dyslipidaemic, and antiplatelet aggregation effects as well as the protective effects against myocardial infarction.

Chapter 1 - Terpenes are a vast group of secondary metabolites to which a great variety of biological properties have been attributed. Among these properties the most widely studied is antimicrobial activity, nevertheless recent researches have demonstrated that terpenes possess also antioxidant, anti-inflammatory and anti-cancer properties. Due to this, terpenes have become of interest in both pharmaceutical and food industries. Considering that plants produce and store thousands of terpenes, plant-derived terpenes represent an alternative to obtain these compounds. Among the vegetable species, oregano is considered one of the most important in terpenes extraction. Terpenes in oregano, mainly represented by the essential oils, are extensively documented for their antimicrobial, antifungal, antiviral, antioxidant, anti-inflammatory and anticancer properties. The aim of this chapter is to describe the biological properties of terpenes from oregano and their mechanism of action.

Chapter 2 - Monoterpenes are renewable resources that are widely used in the pharmaceutical, cosmetic and food industries as active components of drugs and ingredients of artificial flavours and fragrances. Acids catalyzed of terpenes are important synthetic routes to valuable compounds. Traditionally, strong homogeneous acid catalysts, e.g., sulphuric acid, have been used in acid reactions. However, solid acids have many advantages over liquid acids. They do less harm to the environment and do not have corrosion or effluent disposal problems. They are reusable and easy to be separated from liquid products. This review focuses on solid acid catalysts as potential heterogeneous catalysts for terpenes conversion into fragrances and flavours, replacing homogenous ones. Different reactions, such as hydration, alkoxylation, esterification, isomerization and cyclization of terpenes over heterogenous catalyts will be reviewed.

Chapter 3 - Turpentine, which is composed of the volatile terpenes fraction from conifer wood, is the most widely produced essential oil. Of the 330 000 T of turpentine produced worldwide every year, more than half comes from kraft pulp mills. This sulfate turpentine is recovered at a rate of 2-15 kg/T of air-dried pulp, and is composed mainly of 45-95% α-pinene and 0.5-28% β-pinene. The monoterpene recovery rate and composition depend essentially on the wood species used for wood-pulp manufacturing (high terpene content in pine species), but also on the wood processing operations prior to pulping. Avoiding long storage times of wood raw materials prior to pulping will lead to a higher turpentine yield. An exhaustive process-line assessment could help a kraft pulp mill improve its recovery yield. Highly volatile terpene emissions also occur during thermal/mechanical processes in sawmills and thermomechanical pulp (TMP) mills. Between 0.4-1 kg of turpentine per ton of TMP could be recovered in a TMP mill, thereby reducing volatile organic compound (VOC) emissions accordingly. Like the gum turpentine obtained from tree tapping, TMP turpentine can be marketed at a price four to five times higher than that of sulfate turpentine from kraft pulp contaminated with malodorous sulfurous compounds.

Chapter 4 - Cardiovascular diseases remain a global burden with tendency to increase due to behavioural risk factors including tobacco and alcohol abuse, unhealthy diet, obesity, and physical inactivity. Current treatments include clinical procedures and medications that are clearly not sufficient, thus justifying novel strategies that may include preventive approaches, such as a healthy lifestyle. Indeed, it is widely accepted that a Mediterranean-style diet, rich in plant intake, can have cardioprotective effects by attenuating cardiac risk factors and/or modulating disease's onset and progression. In fact, several compounds present in plants, namely secondary metabolites, such as terpenes, have broad therapeutic effects with several studies ascribing beneficial cardiac effects. Bearing in mind that plant essential oils are complex mixtures of several compounds, mainly terpenes of low molecular weight (monoterpenes and sesquiterpenes), the present review focuses on the effects of terpene-rich essential oils. Although several studies describe the cardioprotective effects of these essential oils, a systematic review is lacking. Therefore, a compilation of these studies is presented with special focus on the hypotensive, vasorelaxant, modulation of ion channels, anti-dyslipidaemic, and antiplatelet aggregation effects as well as the protective effects against myocardial infarction. Moreover, putative mechanisms of action involved in these effects are presented as well as an attempt to establish a composition-activity relationship.

In: Terpenes
Editor: Albin Bjarke
ISBN: 978-1-53613-622-7

Chapter 1

TERPENES IN OREGANO: CONSTITUENTS, EXTRACTION, ANALYSIS AND BIOLOGICAL PROPERTIES

Nayely Leyva-López, Erick P. Gutiérrez-Grijalva, Gabriela Vazquez-Olivo, Laura A. Contreras-Angulo, Alexis Emus-Medina and J. Basilio Heredia*

Laboratory of Functional Foods and Nutraceuticals.
Centro de Investigación en Alimentación y Desarrollo,
A.C. Culiacán, Sinaloa, Mexico
CONACYT-Laboratory of Nutrition and Food Plant.
Centro de Investigación en Alimentación y Desarrollo,
A.C. Mazatlán, Sinaloa, Mexico

ABSTRACT

Terpenes are a vast group of secondary metabolites to which a great variety of biological properties have been attributed. Among these properties the most widely studied is antimicrobial activity, nevertheless recent researches have demonstrated that terpenes possess also

* Corresponding Author Email: jbheredia@ciad.mx.

antioxidant, anti-inflammatory and anti-cancer properties. Due to this, terpenes have become of interest in both pharmaceutical and food industries. Considering that plants produce and store thousands of terpenes, plant-derived terpenes represent an alternative to obtain these compounds. Among the vegetable species, oregano is considered one of the most important in terpenes extraction. Terpenes in oregano, mainly represented by the essential oils, are extensively documented for their antimicrobial, antifungal, antiviral, antioxidant, anti-inflammatory and anticancer properties. The aim of this chapter is to describe the biological properties of terpenes from oregano and their mechanism of action.

Keywords: essential oils, *origanum vulgare*, chemical composition, health benefits

1. Introduction

Oregano is the designation used to refer to a great diversity of plants that share a particular flavor and odor (S. E. Kintzios, 2012). At least 61 species and 17 genera belonging to six different botanical families are known as oregano. Verbenaceae and Lamiaceae are the most conspicuous families due to their economic importance. Within the Lamiaceae family are the plants belonging to the genera *Origanum* and *Hedeoma*; while the genera *Lippia* and *Lantana* belong to the Verbenaceae family. The other families are Rubiaceae, Apiaceae and Asteraceae (Baser & Buchbauer, 2015; S. E. Kintzios, 2012). *Hedeoma patens*, *Lippia graveolens*, *Lippia palmeri*, *Lippia alba*, *Origanum dictamnus*, *Origanum hirtum*, *Origanum onites* and *Origanum vulgare* are some examples of oregano species (Baser & Buchbauer, 2015). Oregano species have been used traditionally to treat different diseases and recognized to possess antioxidant and anti-inflammatory properties (Laferriere et al., 1991; Rubió et al., 2013). These properties have been attributed to the presence of secondary metabolites in oregano (Kogiannou et al., 2013; Loizzo et al., 2009).

Different types of secondary metabolites are present in oregano species, such as, terpenes, flavonoids and phenolic acids (Leyva-López et al., 2016). Terpenes are one of the main phytochemicals, with different classes including monoterpenoids, sesquiterpenoids, diterpenoids and triterpenoids being isolated and characterized in these herbs. Several studies have shown that terpenes possess antibacterial, antiviral, antioxidant, anti-inflammatory and anticancer activities. Therefore this chapter is intended to provide an overview of the chemical identities and the pharmacological effects of the terpenes isolated from oregano species; the extraction and analysis methods will also be mentioned here. This information could be useful as a reference for further research.

2. Terpene Constituents of Oregano

Terpenes are substances that are formed by the condensation of subunits of isoprene (C_5H_8), so their basic molecular formula is $(C_5H_8)_n$. These compounds are classified according to their number of isoprene units (Kirby & Keasling, 2009). For instance, terpenes conformed by two units of isoprene are known as monoterpenoids, those with three units of isoprene are named sesquiterpenoids, those with four units are named diterpenoids and so on. Table 2.1 shows the different types of terpenes according with their skeleton.

Table 2.1. Classes of terpenes

Terpene	Number of isoprene units	Number of carbon units
Hemiterpenoids	1	5
Monoterpenoids	2	10
Sesquiterpenoids	3	15
Diterpenoids	4	20
Triterpenoids	6	30
Tetraterpenoids	8	40
Polyterpenoids	>8	>40

2.1. Monoterpenoids and Sesquiterpenoids in Oregano Species

Monoterpenoids and sesquiterpenoids are the main constituents of the essential oils of oregano. Countless studies have been realized to elucidate the chemical composition of volatile components in species such as *Origanum vulgare* spp, *Origanum onites*, *Origanum majorana* and some *Lippia* spp., namely *Lippia graveolens*, *Lippia palmeri* and *Lippia alba*. The monoterpenoids most commonly and abundantly found in oregano species are thymol and carvacrol, but *p*-cymene and γ-terpinene are also frequently present in the essential oils of these herbs (Betancourt et al., 2012; Figueredo et al., 2012; Ocak et al., 2012; Olmedo et al., 2014). Other monoterpenoids found are carvacrol acetate, cymene, linalool, myrcene, *trans*-piperitol, among others. On the other hand, the sesquiterpenes reported to be found in oregano species are *trans*-sabinene hydrate, β-caryophyllene, cadinene, germacrene D and bisabolene (Al Hafi et al., 2016; Leyva-López et al., 2016; Morshedloo et al., 2017) (Table 2.2). It is very important to mention that the terpene profile in oregano will vary depending on several factors, such as, species, growth and climatic conditions, harvest season and geographical location, among others (Asensio et al., 2015; Azizi et al., 2009; Baranauskienė et al., 2013; Morshedloo et al., 2017).

2.2. Diterpenoids and Triterpenoids Found in Oregano Species

Diterpenoids are not very commonly found in oregano species. Carnosol and carnosic acid have recently been identified in extracts from *Origanum vulgare*, *Origanum marjorana* and *Lippia graveolens* (Bower et al., 2014; Bunghez et al., 2015; Mohammad B. Hossain et al., 2011) (Table 2). There are other diterpenoids, such as akhdarenol, akhdardiol, akhdartriol and isoakhdartriol that have been found in *Origanum akhdarense* and *Origanum pampaninii*, but these are very rare and not often found in *O. vulgare* or other *Origanum* species (Spiridon E Kintzios, 2003). On the other hand, the major triterpenoids found in oregano are ursolic, oleanolic and betulinic acid. These three terpenoids have been isolated from aerial parts of *Origanum onites* and *Origanum vulgare* (Baranauskaitė et al., 2016; Sowa et al., 2014; Tepe et al., 2016).

3. Extraction and Analysis of Terpenes from Oregano

3.1. Methods for Extraction of Terpenes from Oregano

The oregano is an aromatic plant with antioxidant and anti-inflammatory properties, attributed to its content of phenolic acids, flavonoids and terpenes (Leyva-López et al., 2016). These last are generally hydrophobic compounds, soluble in organic solvents, liquid, with fragrance, without color, lighter than water and boiling point between 150 and 180°C. Most of terpenes found in oregano spices are volatile in steam, refractive and optically active, only a few of them are solid. These properties, as well as their structure, are determinant in the extraction method to use (Jiang et al., 2016). Among the most commonly used methods are hydrodistillation, solvent extraction and extraction with supercritical fluids (Ludwiczuk et al., 2017; Yadav et al., 2014). In general, the methods used to extract terpenes include the following steps: i) release of its chemical constituents by breaking the plant cell, ii) extraction of the components with solvents or by distillation, iii) separation of the desired terpenes from unwanted contents to avoid confusion in the analysis and iv) analysis of terpenes by an appropriate method (this may vary depending on the polarity and size of the terpenes extracted) (Jiang et al., 2016). Next, the main methods for terpene extraction from oregano will be described.

3.1.1. Hydrodistillation

This is, perhaps the simplest and oldest method used for extraction of the terpene constituents of oregano species. Essentially, the plant, or part of the plant, is placed in boiling water or heated with steam. The applied heat is the main cause of the breakdown of the cellular structure, which results in the release of the aromatic compounds (Tongnuanchan & Benjakul, 2014).

Hydro-distillation is a conventional method where saturated steam is used to obtain essential oils, which are a mixture of mostly terpenes (90-95% monoterpenoids and sesquiterpenoids). In this method the essential oils are evaporated by heating and mixed with water, followed by a liquefaction of the steam in a condenser (Figure 1a). The essential oil remains on the surface of the water, to be subsequently decanted or dried with anhydrous sodium sulfate. Water acts as a barrier protecting the essential oils from overheating, and has the advantage that it is possible to distill at temperatures below 100°C at atmospheric pressure. The determining factors in the quality of the essential oil, apart from the oregano species, is the time, temperature and pressure, in addition to the quality of the equipment used to distill (R. Kumar & Tripathi, 2011; Rassem et al., 2016; Tongnuanchan & Benjakul, 2014). To performance the extraction by hydrodistillation different systems can be used, however the most common is the Clevenger apparatus (Figure 1b). This apparatus consists of three parts, a flat bottom ball flask, in which the material containing the volatile compounds and water, this is connected to a separator, where the essential oil of the distillate is automatically separated in a graduated tube (the Clevenger trap) and a condenser (Clevenger, 1928).

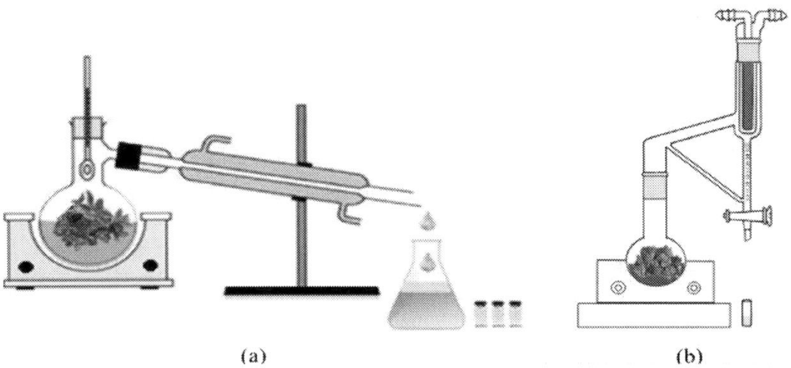

Figure 1. (a) Scheme of the extraction system used in hydrodistillation.
(b) Clevenger apparatus.

Table 2.2. Main terpenes found in some *Origanum* and *Lippia* species

Constituents	*Origanum vulgare* L.	*Origanum majorana*	*Origanum onites*	*Lippia graveolens*	*Lippia palmeri*
Monoterpenoids					
Carvacrol	+		+	+	
Timol	+		+	+	+
p-Cymene	+	+	+		
γ-Terpinene	+	+	+	+	
Terpinen-4-ol	+	+			
Sesquiterpenoids					
trans-Sabinene hydrate	+	+			
β-Caryophyllene	+		+		
Cadinene				+	+
Germacrene D	+				
Bisabolene	+				
Diterpenoids					
Carnosol	+	+		+	
Carnosic acid	+	+	+		
Triterpenoids					
Ursolic acid	+	+			
Oleanolic acid	+	+			

There are several studies where the extraction of essential oils of oregano using hydrodistillation is the main goal. For instance, G. Ozkan et al. (2010) evaluated the composition of the essential oil of leaves from *O. onites* according to different harvest time and they found as carvacrol as the main component (83.97-88.65 %). Thymol (0.00-7.48 %), γ-terpinene (2.63–6.15%), *p*-cymene (1.52–3.16%), α-pinene (0.00–1.88%) and α-terpinene (0.00–1.05%) were also present. In addition, oregano leaves from *Lippia palmeri* S. Wats were distilled using a Clevenger apparatus during 4 h to obtain essential oil and analyzed by GC-MS using a chromatograph coupled to ion trap mass spectrometer, they identified *p*-cymene (22.37%), thymol (21.39%), carvacrol (8.76%) and *p*-thymol (5.94%) (Ortega-Nieblas et al., 2011). Furthermore, Missopolinou et al. (2012) evaluated the composition of essential oils of oregano from different regions in Greece obtained by hydrodistillation for 60-70 min. These authors found that carvacrol was the major constituent in the oregano of all regions. Nevertheless, other terpenes were identified but their content varied depending on the region. Bonfanti et al. (2012) identified the profile of terpenes extracted by hydrodistillation (50-100 g for 3 h) in five samples of oregano (*O. vulgare* ssp *hirtum*) cultivated in different areas of Sicily. In this study different yields and terpenes extracted according to the origin of the oregano were obtained. Monoterpenes and oxygenated monoterpenes were the main constituents of this oregano specie. Begnini et al. (2014), used a Clevenger apparatus to obtain essential oils of *O. vulgare* leaves. The hydrodistillation process lasted 4 h and the distilled was collected every 5 min. The yield of recuperation was 1.33 ± 0.16% of essential oil where 25 terpenes were identified. Terpinen-4-ol was the main component, followed by thymol, γ-terpinene and carvacrol. In addition, Grondona et al. (2014) obtained essential oils from *O. vulgare* ssp. *hirtum* by the same method. In this study the major monoterpene found was γ-terpinene, followed by terpinen-4-ol, α-terpinene, *trans*-sabinene hydrate and thymol among others. On the other hand Olmedo et al. (2014) used hydrodistillation for 1 h in the Clevenger apparatus to obtain terpene extracts from *O. vulgare* leaves. Six major terpene components were found: terpinen-4-ol, carvacrol, linalool, γ-terpinene, α-terpinene and α-terpineol. In addition, Acevedo et al. (2013) evaluated the same species but collected from a different region. In this study was found that the major terpene was thymol (67.51%) (oxygenated compound), followed by *p*-cymene (11.66%), γ-terpinene (5.51%), caryophyllene (5.38%), caryophyllene oxide (2.22%), *trans*-α-bergamoteno (1.65%), eugenol (1.49%) and α-bergamoteno (1.32%). Also, Chishti (2016), collected the same species (*O. vulgare*) in three regions of Kashmir Himalaya and found that the terpene γ-terpinene (monoterpenic hydrocarbon) was the highest in the oregano collected from Srinagar and Pulwama, and *p*-cymene (21.13% and 33.29 %) was the main component in the ones collected in Pulwama and Tangmarg, respectively. In addition they identified moderately high percentages of sabinene and carvacrol. F. Han et al. (2017) evaluated the antioxidant activity of essential oils from different parts (leaves-flowers, steam and roots) of *O. vulgare*, they used Clevenger apparatus to extract the essential oil and gas chromatography-mass spectrometry (GC-MS) to identified the components of the oil. They identified 59 compounds from the three different parts of oregano. The main constituents were carvacrol, thymol and caryophyllin, which were present in all the parts of oregano.

All the studies carried out on oregano show differences in the terpene content according to the amount of sample used, the origin of the oregano, extraction times and the method of identification and quantification.

3.1.2. Extraction Using Organic Solvents

The extraction using organic solvents is a liquid-liquid extraction used with two liquids immiscible and it is based on the solubility of the compounds of interest (Rassem et al., 2016). To carry out this method the solvent comes in contact with the plant material and dissolves the substances, then evaporate and concentrate at low temperature to eliminate the solvent and to obtain the desired compounds. Different solvents have been used such as, acetone, hexane, petroleum ether, chloroform, ethyl acetate and methanol or ethanol. Nonetheless the ideal solvent does not exist, it is important for its selection to consider the ability to rapidly dissolve all compounds, the boiling point for its rapid elimination and avoid losses, and that is inert so it does not react with the components of interest. Essential oils, sesquiterpenes, diterpenes, sterols and less polar triterpenes can be extracted with petroleum ether or chloroform; oxygenated diterpenes, sterols and triterpenes with ethyl acetate; while highly oxygenated triterpenes can be obtained with ethanol, methanol and water. A protocol to achieve a total extraction of plant material is to first use a polar solvent such as acetone, aqueous methanol (80%) or ethanol followed by a reextraction with hexane, chloroform and ethyl acetate, to be able to extract terpenoids and sterols. This method produces high quantities of essential oils, it is relatively simple and efficient. However, it has disadvantages such as the extraction time, relatively high solvent consumption and poor reproducibility, in addition to leaving inherent residues (Grández & Einstein, 2017; Ludwiczuk et al., 2017; Rassem et al., 2016; Tongnuanchan & Benjakul, 2014; Velasco et al., 2007) (Table 3.1).

Table 3.1. Terpenes identified in oregano species extracts obtained by different organic solvents

Oregano Species	Extraction solvent	Terpenes identified	Reference
H. patens	Chloroform	Thymol, *trans*-piperitol, carvacrol acetate, carvacrol, camphene, β-myrcene, γ-terpinene, cis-p-mentha-1(7), 8-dien-2-ol, α-muurolene, α-calacorene, bulnesol, cadalene, viridiflorol	(Leyva-López et al., 2016)
L. graveolens	Chloroform	Thymol, carvacrol acetate, carvacrol, camphene, β-myrcene, γ-terpinene, cis-p-mentha-1(7),8-dien-2-ol, viridiflorol	
L. palmeri	Chloroform	Thymol, α-cedrene, *trans*-piperitol, eugenol, carvacrol acetate, β-selinene, γ-cadinene, spathulenol	
Oregano (undetermined species)	Methanol 80%	Carnosol, carnosol acid, methyl carnosate.	M. B. Hossain et al. (2010)
O. vulgare L.	Dichlromethane	Carvacrol, γ –terpinene, carvacrol methyl ether, trans- β-Ocimene, β-caryophyllene, ρ –cimene.	Lukas et al. (2008)
	Methanol	Thymol, carvacrol, α-terpinene, α-phellandrene, γ –terpinene, terpinen-4-ol, caryophyllene, germacrene D.	Rodríguez-Solana et al. (2015)
	Ethyl acetate-pentane (2:1)	γ -terpinene, trans-sabinenehydrate, p-cymene, sabinene, cis- and trans-b-ocimene and thymol.	Crocoll et al. (2010)
	n-Hexano β-	Carvacrol, terpinen-4-ol, β-caryophyllene, germacrene D	De Falco et al. (2013).
O. vulgare ssp. *hirtum*	Dichlromethane, 4-methyl-1-pentanol	Carvacrol, thymol, ρ -cimene, γ -terpinene, mircene, β-pinene, β-caryophyllene, α-pinene.	Grevsen et al. (2009)
	Ethyl acetate, ethanol	Thymol, γ –terpinene, p-cymene, α-terpinene.	Tuttolomondo et al. (2013)
O. vulgare ssp. *virens*	n-hexane, cloroform, ethy acetate and methanol	γ-terpinene, thymol, ρ-cymene, β-caryophyllene, germacrene D., cis-b-ocimene.	Castilho et al. (2012)

For a long time this method has been used in the extraction of terpenoid compounds in different plants, including oregano, which is distinguished by its medicinal properties, attributed mainly to its essential oil rich in terpenes.

3.1.3. Supercritical Fluid Extraction (SFE)

This technology has been described as environmentally safe or green technology, since it does not generate pollution using non-toxic organic solvents (Gil-Chávez et al., 2013). This method is based on some properties of fluids such as density, diffusivity, dielectric constant and viscosity (Gil-Chávez et al., 2012). The method consists in passing a supercritical fluid (SF) (extractor solvent) through a chamber that contains the sample (cut or ground) of interest. Then the SF is eliminated by decompression and finally the essential oils are obtained (Pereira & Meireles, 2010).

The use of this technology in the extraction of essential oils of aromatic plants intended for the pharmaceutical, food and cosmetic industries has been of great importance. Essential oils contain hydrocarbon and oxygenated monoterpenes and sesquiterpenes used in different industries, which is why it is imperative to extract them in a clean way without residues. Therefore the use of supercritical fluids, such as CO_2, represents a good alternative for the extraction of terpenes (Capuzzo et al., 2013). Within the species studied, oregano is rich in essential oils and other antioxidant compounds.

There are several studies conducted using SFE in oregano where they used different conditions to extract essential oils. Using this extraction technology with supercritical CO_2, essential oil was obtained from dried and ground oregano (*O. vulgare*) leaves. The pressure and temperature conditions used were 30 MPa and 40°C, respectively. The extract obtained was injected into a gas chromatograph coupled to masses and 16 compounds were detected, the compounds found in greater percentage were: *trans*-sabinene hydrate (45.21%), thymol (24.10%), carvacrol (7.99%) and *ρ*-cymene (5.70%) (Ocana-Fuentes et al., 2010). Furthermore, Ivanovic et al. (2011) investigated the kinetics and mass transfer phenomena occurring during the SFE from plants with different initial composition. They used dried leaves of oregano (*O. vulgare*), the extraction was performed at 10 MPa and 40°C in order to isolate essential oil rich in terpenes. First the flow was kept static to break the storage structures of the essential oil, the glandular trichomes, then the flow was continuous. The extract obtained was analyzed by chromatography/mass spectrometry (GC/MS). The main compounds found were carvacrol and thymol. Santoyo et al. (2014) evaluated terpenes from dried leaves of oregano obtained by SFE. The extraction process was performed at 30 MPa and 39.85°C with a flow rate of CO_2 of 50 g•min^{-1}. The characterization of the oregano extract was carried out by GC-MS and the compounds identified were *trans*-sabinene hydrate (45.21%), thymol (24.10%), carvacrol (7.99%) and *p*-cymene (5.70%). Fornari, Ruiz-Rodriguez, et al. (2012) used SFE with CO_2 to obtain essential oils from *O. vulgare*. The conditions were 30 MPa of pressure and 39.85°C of temperature, while the fluid flow rate was set to 2.4 kg•h^{-1}. The extract was analyzed by GC-MS to determine the essential oil composition. The main compounds identified were thymol (37.83%), *trans*-sabinene hydrate (35.53%) and carvacrol (13.86%). Rodríguez-Solana et al. (2015) also obtained essential oils from *O. vulgare* using SFE with CO_2. The conditions of the process were a flow rate of 40 g•min^{-1}, the pressure and temperature were maintained at 24 MPa and 59.85°C, respectively. These extracts contained thymol (40.77%), terpinen-4-ol (9.68%), carvacrol (6.47%) and caryophyllene (6.15%) as main constituents. In another study by Fornari, Vicente, et al. (2012) the main terpenes found in oregano extracts obtained by SFE (30 MPa and 40°C) were *cis*-sabinene hydrate (38.25%), thymol (35.73%), and carvacrol (11.77%).

3.2. Analysis of Terpenes from Oregano Using Chromatography

As it was mentioned above, oregano is recognized by its great diversity of terpenes, which are considered as one of the most natural and structurally diverse natural products (Ludwiczuk et al., 2017). The analysis of terpenes requires the implementation of techniques that allow their proper identification. Chromatographic techniques are, undoubtedly, the best alternative for terpene analysis. Chromatography is the science that studies the separation of molecules based on their structural differences, complex mixtures of compounds can be separated by a support and the interactions of molecules with that support. In this way the chromatographic separation can be done using different media such as paper (paper chromathography), silica on glass plates (thin-layer chromatography), volatile gases (gas chromathography) and liquids (liquid chromatography) (Chinou, 2010). In this sense, the main techniques that have been used for the study of terpenes in oregano will be briefly described next.

3.2.1. Thin Layer Chromatography

Thin layer chromatography (TLC) is used as a tool to identify and quantify compounds (Meyers & Meyers, 2008). TLC involves the movement of a liquid phase (usually an organic solvent) through a thin layer of stationary phase (usually silica gel, SiO2), supported by a rigid or semi-rigid support, glass, aluminum or plastic, among others (Flanagan et al., 2008). Some advantages of TLC are simplicity, economy of materials, selective detection methods and simultaneous analysis of samples (Dugo et al., 2000). TLC is mainly used for the qualitative identification of different types of compounds such as fatty acids, flavonoids, saponins, alkaloids, polyphenols and terpenes, among others (Alor-Chávez et al., 2017).

The analysis of terpenes by TLC is generally fast, however, it is necessary the use of some reagents such as sulfuric acid, because terpenoids are colorless. Silica plates are mainly used as stationary phase, but alumina and cellulose are also employed for the separation of sesquiterpenes, diterpenes, triterpenes and tetraterpenoids (Ludwiczuk et al., 2017).

Oregano is characterized for being rich in antioxidant compounds, however, the composition of oregano is variable, depending on factors such as the species, subspecies, type of extraction and part of the plant (Bertelli et al., 2003). Terpenes are the main group of compounds present in oregano, therefore various methods have been used for their analysis.

Several researchers have used TLC for the study of terpenes in oregano. For instance, Pensel et al. (2014) analyzed the essential oil of *O. vulgare* by TLC. In this study silica gel plates (0.2 mm Kieselgel 60 F254, Merck) and toluene:ethyl acetate (93:7) were used as stationary and mobile phase, respectively. The separated compounds were treated with sulfuric acid and vanillin and heated to 110°C. Five bands were observed, the superior band was the most intense with an Rf 0.60, this band was related to the presence of thymol and carvacrol. This finding was confirmed by gas chromatography. Likewise, Bozin et al. (2006) identified carvacrol (Rf = 0.62), thymol (Rf = 0.73) and a mixture of mono and sesquiterpene hydrocarbons (Rf = 0.98) in the oil of *O. vulgare* combining gas chromatography and a rapid analysis of TLC with silica gel plates (F254 MercK) and the mobile phase benzene:ethyl acetate (95:5). Additionally, the identification of it thymol and carvacrol, in the varieties *Origanum* x *applii* and *Origanum* x *majoricum* by HTPLC has been reported. In this study plates of silica gel (60 Macherey-Nagel) and toluene were used as stationary and mobile phase, respectively. Thymol was the predominant monoterpenoid in both species (Amadio et al., 2011). In this context, the study of terpenes in oregano by TLC is an economic and simple analytical technique for the rapid confirmation of its presence.

3.2.2. High-Performance Liquid Chromatography Analysis

Although the analysis by high-performance liquid chromathography (HPLC) is not recommended for the analysis of terpenes, several researchers have used this analytical technique for their study (Koch et al., 2010). Adequate extraction is essential for the analysis of monoterpenes, sesquiterpenes, diterpenes and tetraterpenes. In this case the diterpenes of medium to low polarity, have been obtained using solvents like dichloromethane, ethyl acetate, methanol and sometimes more polar solvents such as water is used. This promotes the extraction of soluble diterpenes and flavonoid derivatives (Bunghez et al., 2015). Likewise, non-polar solvents, such as n-hexane, are used for the extraction of diterpenes of low polarity, for their subsequent preparation and resuspension in mobile phase for their analysis (Hajnos, 2010). Besides, HPLC is used more frequently for the determination of triterpenes and steroids, since it can effectively treat highly polar non-volatile compounds (Oleszek & Stochmal, 2010).

For the analysis of terpenes in plant extracts it should be considered that some are characterized by containing two fractions, one of the fractions is the oil constituted mainly by volatile terpenes, while the other fraction represents phenolic compounds including phenolic terpenes (Zabot et al., 2014). In this regard, the chromatographic conditions will depend on the extract obtained. Some researchers have developed techniques for the identification and quantification of terpenoids by liquid chromatography in oregano, with this intention Téllez et al. (2014) proposed a method by HPLC with fluorescence detector for the identification and quantification of thymol and carvacrol in oregano oil. This method showed a detection limit of 0.007 mg/L for thymol and 0.002 mg/L for carvacrol and a maximum recovery of 90.95% and 98.68% for thymol and carvacrol, respectively. This method could be a tool for the rapid determination of quality based on the proportions of thymol and carvacrol of essential oils. The conditions of analysis are presented in table 3.2. Similarly, Zabot et al. (2014) developed and validated a method by Ultra Performance Liquid Chromatography (UPLC) with mass detector for the determination of phenolic terpenes as rosmanol, carnosol, carnosic acid and methyl carnosate, with a total analysis time of 10 min. On the other hand, Bunghez et al. (2015) analyzed hydrophilic extracts of oregano by liquid chromatography coupled to masses with ionization by electrospray and time of flight (LC-ESI-QTOF-MS). The major compounds corresponded to carnosol and carsonic acid, these diterpenes represented more than 14% of peak area fractions. Likewise, M. B. Hossain et al. (2010) characterized the phenolic composition of five extracts from species belonging to the Lamiaceae using liquid chromatography coupled to masses with electrospray ionization (LC-ESI-MS/ MS), identifying 30 phenolic compounds and four terpenes in oregano (methyl carnosate, methoxycarnosol, carnosol and carnosic acid).

3.2.3. Gas Chromatography Analysis

Gas chromatography (GC) is a separation technique that involves a stationary phase and a mobile gas phase. In the separation by GC, the sample is vaporized and transported by a mobile gas phase (carrier gas) through a column, the separation of the different compounds is achieved as a function of the relative vapor pressure and affinity of the stationary phase (Piantanida & Barron, 2014). Terpenes present closely related physical and chemical characteristics, which complicates their analysis. Most of the terpenes in oregano, with some exceptions, are volatile compounds; therefore, gas chromatography can be an appropriate option for their study. If gas chromatography is combined with mass spectrometry, highly complex samples could be determined (Witte, 1986).

Calvo-Irabién et al. (2014) identified and quantify terpene compounds of oregano (*Lippia graveolens* Kunth) by GC coupled to masses (MS) and GC-FID (Flame ionization detector), respectively. These authors identified 70 compounds, which correspond to 89-99% of the total composition of the oils. Most of the compounds were oxygenated monoterpenes and oxygenated sesquiterpenes. In other investigations, essential oils of oregano (*O. vulgare* L.) obtained by solvent-free microwave extraction (SFME) were analyzed by mass-coupled gas chromatography (GC-MS). It was observed that the essential oil was composed of oxygenated compounds (80-85%), hydrocarbon monoterpenes (13-16%) and sesquiterpenes (1-1.6%). The major compounds were thymol, *p*-cymene, carvacrol, γ-terpinene, β-myrcene and α-terpinene (Bayramoglu et al., 2008). Likewise, the composition of the essential oil of oregano in two locations in Mexico (Álamos and Puerto del Oregano) were analyzed by GC-MS. In this study, 50 and 60 compounds were identified in the oil of oregano collected in Álamos and Puerto del Oregano, respectively. More than 97% of the essential oils was composed by monoterpenes and sesquiterpenes for oregano from both localities. The main compounds in the essential oil of Alamos were *p*-cymene (22.3%), followed by thymol (21.3%), iso-aromandrene (16.7 %), carvacrol (8.8%), γ-terpinene (6.7%) and *p*-thymol (5.9%). While in the essential oil of oregano from Puerto del oregano corresponded to carvacrol (24.5%), thymol (15.1%), *p*-cymene (14.2%), caryophyllene (9.9%), thymol acetate (5.5%) and α-bisabolene (4.8%) (Ortega-Nieblas et al., 2011).

Similarly, Rodriguez-Garcia et al. (2016) evaluated the composition of essential oils from *L. graveolens* using GC. Carvacrol was the main component (47.4%), followed by *p*-cymene (26.4%), thymol (3%) and caryophyllene (1.3%). Furthermore, Peschiutta et al. (2016) identified 37 compounds in the essential oil of *L. graveolens* using GC-MS and found finding the same major terpenes but in different proportions for thymol (22.8%), carvacrol (22.7%) and *p*-cymene (18.8%). Likewise, the evaluation of essential oil of *L. graveolens* obtained by Steam Distillation and analyzed by GC-MS showed that the major compounds correspond to carvacrol, thymol, and β-cymene (Soto-Armenta et al., 2017). On the other hand, Leyva-López et al. (2016) analyzed, by GS-MS, the terpenoid profile of extracts of the three main species of oregano marketed in Mexico (*L. graveolens, L. palmeri and H. patens*), identifying 19 compounds in *H. patens*, while in the genus *Lippia* only eight compounds were identified. In general, the major compounds correspond to monoterpenes and sesquiterpenes. Thymol and carvacrol acetate were found in the three species. Table 3.3 shows the chromatographic conditions of GC used by various investigators in the identification of terpenes in oregano. Based on the above, we can conclude that GC is an adequate technique for the identification of terpenoids in samples of oregano.

4. Biological Properties of Terpenes from Oregano

4.1. Antimicrobial Properties of Terpenes from Oregano Species

As a result of the emergence of antibitoic resistant strains and increased incidences of drug resistant biofilm associated infections; there has been an ongoing effort in searching for new alternatives to fight these infections (Adame-Gallegos et al., 2016). One of the most important alternative to antibiotics to multidrug-resistant bacterial strains is the use of terpenes from different sources, from which herbs and spices are the leading target for their terpene-rich content.

Table 3.2. Conditions for the identification of terpenes of oregano using liquid chromatography

Sample	Main Compounds	Stationary phase	Mobile phase	Detection	Reference
Turkish oregano	Methyl carnosate, Methoxy carnosol, Carnosol and Carnosic acid	Atlantis T3 C18 column (Waters Corp., Milford, MA; 100 mm x 2.1 mm; 3 µm)	A-5% aqueous formic acid B-0.5% formic acid in (50:50) acetonitrile:methanol Gradient elution	ESI-MS/MS	M. B. Hossain et al. (2010)
O. vulgare	Carnosol and Carnosic acid	Thermo Scientific Acclaim C18 column (50 mm x 2.1; 3µm)	A- 0.1% aqueous formic acid B- 0.1% formic acid in acetonitrile Gradient elution	ESI-QTOF-MS	Bunghez et al. (2015)
	Carvacrol and Thymol	Purospher STAR rp-18e Column (150 mm x 4.6mm; 5 µm)	Water-acetonitrile (50:50, v/v), isocratic elution	FL- 230 excitation and 280 emission	Téllez et al. (2014)
Salvia officinalis *Rosmarinus officinalis* *O. vulgare* Mixture of herbs (sage,oregano, thyme and rosemary) Mixture for chimichurri sauce (onion, garlic, rosemary, sage and red pepper)	Rosmarinic acid, Rosmanol, Carnosol, Carnosic acid and Methyl carnosate	Acquity, UPLC BEH C18 Colum (50 mm x 2.1mm;1.7 µm)	A- Water with 0.1% (v/v) of acetic acid B- Acetonitrile with 0.1% (v/v) of acetic acid	PDA MS single quadrupole	Zabot et al. (2014)

Table 3.3. Conditions for the identification of terpenes of oregano using gas chromatography

Sample	Main Compounds	Stationary phase/column	Temperature program	Detection	Reference
L. graveolens, L. palmeri, H. patens	Carvacrol acetate and thymol	Agilent Technologies Ultra 1 column (25 m × 0.32 mm; 0.52 µm)	75°C for 4 min, then 75°C to 200°C at 5°C/ min, and held for 200 °C for 1 min.	GC-MS	Leyva-López et al. (2016)
L. palmeri S. WATS	*p*-Cymene, thymol, iso-aromandrene, carvacrol, γ-terpinene and caryophyllene	Column DB-5MS (30m x 0.25 mm; 0.25 µm)	50°C for 10 min, then 290°C with 5°C/min and hold at 290°C for 10 min.	GC-MS Retention times	Ortega-Nieblas et al. (2011)
L. graveolens	Carvacrol, p-cymene, *β*-caryophyllene oxide and eucalyptol	Column DB-5 (10m x 0.18 mm; 0.18 µm)	From 40 to 260°C at 4°C/min, held for 20 min, and raised to 340°C at 4°C/ min.	GC-MS-TOF	Rivero-Cruz et al. (2011)
	Carvacrol and thymol	Agilent Technologies Ultra 1 column (25 m x 0.32 mm; 0.52 µm)	75°C for 4 min, then 75°C to 200°C at 5°C/ min, and held for 200°C for 1 min.	GC-MS	Calvo-Irabién et al. (2014)
	Carvacrol, *p*-cymene, thymol and caryophyllene.	Column DB-5 (30 m × 0.25 mm; 0.25 µm).	From 55 to 65°C with 1°C/min, and held for 3min, then 290°C with 10°C/min. and held 290°C during 10 min.	GC-MS	Rodriguez-Garcia et al. (2016)
	Thymol, carvacrol and p-cymene	DB-5 capillary column	60°C/5 min, then 60°C to 170°C at 4°C/min, and 240°C at 20°C/min	GC-MS	Peschiutta et al. (2016)
	Carvacrol, thymol, and β-cymene	HP-5 MS (30 m x 250 µm; 0.25 µm)	50°C/3 min, then 50°C to 250°C at 20°C/min, and held for 250°C/10 min.	GC-MS	Soto-Armenta et al. (2017)
O. vulgare L.	Thymol, *p*-cymene, carvacrol, γ-terpinene, β-myrcene and α-terpinene	Column HP-5MS (30m x 0.25mm; 0.25 µm).	From 50°C to 150°C with 3°C/min, held at 150°C/10 min, to 250°C 10°C/min.	GC-MS Retention times	Bayramoglu et al. (2008)

In particular, the antimicrobial properties of terpenes from different oregano species have been the focus of several studies (Adame-Gallegos et al., 2016; Burt, 2004; Jayasena & Jo, 2013; Rodriguez-Garcia et al., 2016). It has been generally suggested that the antimicrobial mechanism of action of plant terpenes has been suggested through cell membrane damage, which causes increased cell permeability affecting membrane permeability for ions and protons, affects ATP production, interfere on protein synthesis, cause cellular pH disturbance, induce cytoplasmic changes and interfere with quorum sensing (Scandorieiro et al., 2016).

Origanum vulgare is one of the most known species of oregano worldwide. And, is commonly used in studies regarding antimicrobial properties of terpenes. For instance, *O. vulgare* has shown to be effective in inhibiting *Salmonella enterica* and *Listeria monocytogenes* strains. In this regard, Mazzarrino et al. (2015) showed that the bactericidal effect on *Salmonella* and *Listeria* strains was concentration-depdendant, with higher antimicrobial activity at highest terpene concentrations. Moreover, they suggest that due to the mixed action as antimicrobial and antioxidant, terpenes from *O. vulgare* are a promising alternative as multifunctional biopreservatives. Similarly, it has been reported that terpenes from oregano are more effective inhibiting Gram-positive bacteria. In this regard, Grondona et al. (2014) evaluated terpenes from *Origanum vulgare* against Gram-positive (*Staphylococcus aureus*, *Streptococcus faecalis*, *Streptococcus pyogenes*, and *L. monocytogenes*) and Gram-negative (*Salmonella* spp. and *Shigella* spp.) bacterial strains. They found that the maximum inhibitory activity was observed against *S. aureus*, and *Shigella*. *Origanum vulgare* terpenes have also been effective against the Gram-positive: *S. aureus*, *Bacillus cereus*, *Enterococcus fecalis*, and *L. monocytogenes*; and Gram-negative: *Escherichia coli*, *Yersinia enterocolitica*, *Salmonella cholerasuis*, and *Pseudomonas aeruginosa* bacteria (López et al., 2007).

Although carvacrol has been considered as the main antimicrobial agent from oregano terpenoids, Busatta et al. (2007) attributed terpinen-4-ol as the main antimicrobial agent on the the antimicrobial activity of commercial Chilean oregano against Gram-positive (*E. faecalis*, *S. aureus*,

Streptococcus mutans) and Gram-negative (*Aeromonas sp, E. coli, Klebsiella pneumonae, Salmonella choleraesius, Serratia sp, Shigella flexneri*) inoculated in fresh sausage. This was hipothezised as it was the main component with 21.43% of the total, only followed by γ-terpinene (12.32%), carvacrol (11.67%) and thymol (9.45%). On the other hand, Rivero-Cruz et al. (2011) showed the antibacterial properties of *Poliomintha longiflora* and *L. graveolens* against *Bacillus subtilis, S. aureus, E. coli, Salmonella typhi* and *Pseudomonas aeruginosa*. Interestingly, terepenes from *P. longiflora* had higher antimicrobial activity than *L. graveolens* against *S. aureus* (MIC 128 μg•mL^{-1}) and *B. subtillis* (MIC 128 μg•mL^{-1}), this has been related to the higher carvacrol content found in *P. longiflora*. Another study attributes the antibacterial properties of *O. vulgare* to its high content of carvacrol, thymol and eugenol (Ashraf et al., 2017). Likewise, O. E. Ozkan et al. (2017) showed that the main constituent of terpene extracts from *O. vulgare* ssp. *hirtum* were carvacrol (63.97%), *p*-cymene (12.63%) and linalool (3.67%). And the high content of carvacrol might be related to the inhibition of the extracts against *B. subtilis, C. albicans, E. faecalis, E. aerogenes, E. durans, E. faecium, E. coli, K. pneumoniae, L. monocytogenes, L. innocua, P. aeruginosa, P. fluorescence, S. infantis, S. kentucky, S. typhimurium, S. aureus* and *S. epidermidis* with a minimal inhibitory concentration value of $\leq$ 0.195 μg•mL^{-1} for all microorganisms. S. M. B. Hashemi et al. (2017) showed that terpenes from *O. vulgare* extracted by ohmic heating-assisted hydrodistillation had no effect on the antibacterial properties of the tested samples. And that *O. vulgare* terpenes were more efective inhibiting the growth of Gram-positive bacteria than Gram-negative ones; where a better inhibitory activity was found against *S. aureus* and *B. cereus* (MIC values around 6.25 μg•mL^{-1}) than *P. aeruginosa* (MIC values of 25 μg•mL^{-1}). This effect was attributed to the main constituents that were thymol and terpinen-4-ol.

Moreover, the phenological stage (vegetative, flowering or post-flowering) in which terpenes are extracted from oregano is related to the chemical composition of terpene-yielding during extraction, which can be reflected on their antimicrobial properties (Bouyahya et al., 2017; Sellami et al., 2009). In this subject, Bouyahya et al. (2017) showed that *Origanum compactum* terpenes antibacterial activity against *S. aureus* MBLA is best at flowering stages; whilst at vegetative and flowering stages higher inhibition of *Proteus mirabilis* was higher.

Foodborne microorganisms are the cause of several illnesses. Chlorine-based products are the main used desinfectants. However, chlorine use may lead to the formation of carcinogenic chlorinated compounds in water, thus there is need to find new decontamination alternatives (Gündüz et al., 2010). Terpenes from herbs and spices have been proposed as an alternative to be used in foodstuff due to their proven antimicrobial properties (Rodriguez-Garcia et al., 2016). The use of oregano terpenes is often limited due to organoleptic considerations, since effective antimicrobial doses may exceed organoleptically acceptable doses in food products. As a result, some studies have used a combination of oregano terpenes along with a common desinfectant. For example, terpenes from *O. onites* at a concentration of 75 ppm along with 50 ppm of chlorine on lettuce leaves were effective against *S. typhimurium* (Gündüz et al., 2010). Terpenes from oregano have also been incorporated into meat to inhibit the growth of bacteria during the meat drying process; with an effectiveness against *S. enteriditis* and *E. coli* with MIC values of 0.038 $mL{\bullet}L^{-1}$ and 0.028 $mL{\bullet}L^{-1}$ respectively. However, lower concentrations of terpenes are suggested as they resulted to be more acceptable for consumption by the sensory quality of the dried meat (Hernández et al., 2017). Likewise, Van Haute et al. (2016) achieved inhibition of *E. coli* by incorporating 3% of *Origanum compactum* terpenoid-extract into marinades used on pork filet and salmon, with an acceptable odor after 24 h of storage. The authors discuss the lack of use of a method that determines the active dose. Therefore propper comparision of research regarding antimicrobial properties of terpenes in foodstuff is hampered.

Incorporation of oregano terpenoids into edible films is of interest since these molecules are prone to exidation by light and air, and also as a way to diminish its sensorial effect on foodstuff. For instance, Seyed Mohammad Bagher Hashemi and Mousavi Khaneghah (2017) incorporated terpene extracts from *O. vulgare* subsp. *viride* into edible films from basil seed gum. The main constituents of the sample were thymol (32%), followed by 4-terpineol (12.10%), α-terpinene (10%), trans sabinene hydrate (6.1%), carvacrol (6.0%), sabinene (5.0%), and p-cymene (3.7%). The authors found that basil seed film enriched with 6% of terpenes from oregano resulted in the inhibition of *S. typhimurium*, *E. coli* O157:H7, *P. aeruginosa*, *S. aureus*, and *B. cereus*, and a good sensory perception. Govaris et al. (2010) studied the antimicrobial combined effect of oregano terpenoids from *O. vulgare* subs. *hirtum* at a concentration of 0.6 and 0.9% with nisin at 500 and 1000 $IU \cdot g^{-1}$ against *S. enteritidis* on minced sheep meat. They found that inclusion of terpenes at the tested concentrations was organoleptically acceptable; and that the combination of 0.6% terpenes and nisin at 500 $IU \cdot g^{-1}$ was more effective against *S. enteriditis*. Moreover, nanoemulstions of oregano terpenes is another method studied to diminish their effect on the sensorial properties of foodstuff. On this subject, Moraes-Lovison et al. (2017) encapsulated *O. vulgare* terpenes in nanoemulsions as a way to cut down their effect on the sensorial properties of chicken pâté but preserving its antimicrobial properties. The authors reported antimicrobial properties against *S. aureus* (ATCC 25923) and *E. coli* (ATCC 25922), with MIC values of 0.56 and 0.60 (mg oregano essential oil$\cdot mL^{-1}$). However, their main finding was that the nanoemulsification of oregano terpenes did not affect their antimicrobial properties as compared to pure terpene-extracts. Thus, the incorporation of oregano terpenes into a lipid system would minimize the unfavorable effects on the sensory properties whilst keeping their antimicrobial activity.

As aforementioned, to cut down the effect of terpene addition on the sensorial properties of foodstuf when used as antimicrobial agents, new strategies of inclusion have been studied. Among the most promising is the use of biopolymers with terpene incorporation, where their sensory affection is diminish and the antioxidant stability of terpenes is maintained. Jahed et al. (2017) incorporated terpene extracts from *O. vulgare* ssp. *gracile* in chitosan films; it is important to mention that the major components of this extract were carvacrol, γ-terpinene and *m*-cymene. The antimicrobial properties of chitosan-terpene films were assesed against *E.*

coli (Gram-negative) and *B. cereus* (Gram-positive). Pure chitosan film had antibacterial effect against the studied strains, however, this effect increased when 5% of *O. vulgare* terpene-rich extract was incorporated. This was attributed to the hydrophobic properties of carvacrol in oregano which permit them to disturb the bacterial lipid membrane thus affecting the cell in the aforementioned mechanisms. Likewise, Arana-Sánchez et al. (2010) microencapsulated *L. graveolens* terpenoids with β-cyclodextrin through spray-drying and tested their inhibitory effect against *E. coli*, *P. aeruginosa* and *S. aureus*. They showed that microencapsulated terpenes had antimicrobial activity against *E. coli* and suggested that these are suitable agents to inhibit dangerous pathogens as well as water soluble disinfectants for organic foodstufs.

Oregano terpenes have also been incorporated into edible coatings to be used in ready-to-eat products. For example, Benavides et al. (2012) showed that incorporated 1% terpene extracts from *O. vulgare* into alginate films had antimicrobial properties against Gram-positive (*S. aureus* and *L. monocytogenes*) and Gram-negative (*E. coli* and *S. enteritidis*) bacteria. However, they discussed that the antibacterial activity of pure terpenes from *O. vulgare* was higher than the terpene-alginate films. This effect was attributed to the partial loss of compounds during film preparation; and it may also be due to the slower liberation of terpenes from the alginate matrix, or to the possible interactions between the hydroxyl groups of phenolic terpenes and alginate polymeric chains. Similarly, Emiroğlu et al. (2010) evaluated the effect of the incorporation of terpenes from *O. heracleoticum* at 1, 2, 3, 4 and 5% into soy protein edible films during the preparation of beef patties on the antimicrobial inhibition of *E. coli*, *E. coli* O157:H7, *S. aureus*, *P. aureginosa* and *Lactobacillus plantarum*. The most significantly inhibitied bacteria were *E. coli*, *E. coli* O157:H7 and *S. aureus*. However, the antimicrobial effect was better when terpene-extracts were incorporated directly into beef patties than when used as a biobased edible film-essential oils fromo oregano. Similarly, Jouki et al. (2014) evaluated the effect of the incorporation of commercial oregano terpenes at 1, 1.5, and 2% v/v in quince seed mucigale films on their antibacterial properties against *S. typhimurium*, *P. aeruginosa*, *S. aureus*, *E. coli*,

Shewanella putrefaciens and *Yersinia enterocolitica*. Their results showed that 1% terpene-quince seed mucilage was not effective inhibiting *S. typhimurium* and *P. aeruginosa* but exhibited strong inhibitory effect on the growth of *S. aureus*, *E. coli*, *S. putrefaciens* and *Y. entercolitica*.

Terpenes from oregano have also been recognized for their quick inhibitory response against bacteria. Scandorieiro et al. (2016) reported that *O. vulgare* terpenes acted within 10 mins against *S. aureus* (ATCC 25923), *S. mutans* (ATCC 25175), *E. coli* (ATCC 25922), *S. enterica Enteritidis* (ATCC 13076), *K. pneumoniae* (ATCC 10031), *K. pneumoniae* (ATCC 700603), *S. enterica Typhimurium* UK-1, MRSA N315, MRSA 101, MRSA 107, *E. coli* ESBL 167, *E. coli* ESBL 169, *E. coli* ESBL 176, *E. coli* ESBL 192, *E. coli* KPC 131, *E. Coli* KPC 133, and *A. baumannii* CR 01. Additionally, the time of action was reduced when terpenes were used in combination with silver nanoparticles.

Recently, it has been shown that oregano terpenoids are also active against *Borrelia burgdorferri*, the causative agent of Lyme disease (Feng et al., 2017). The study is particularly interesting, since terpenes showed higher inhibitory activity than daptomycin, a drug commonly used against stationary-phase *Borrelia burgdorferri* persisters; additionally, oregano terpenoids were able to disolve aggregated biofilm-like structures. Nonetheless, further *in vitro* and *in vivo* studies are needed in order to establish safe doses and the pharmacokinetic properties of these compounds before they can be considered as novel alternative in the treatment of persistent any pathogen-caused disease.

4.2. Antifungal Action of Terpenes from Oregano

Terpenes from oregano have also been tested as antifungal agents. Nonetheless, there are fewer studies[1] on this subject than on their antibacterial properties. As it has been previously mentioned, the most studied oregano species is *O. vulgare*. On this sense, terpene-rich extracts from *O. vulgare* have shown antifungal properties against *Mucro*

[1] As literature search by the authors.

racemosus, *Aspergillus fumigatus*, *Cladosporium sphaerosperum*, *Acremonium strictum*, and *Aspergillus niger*; which has attracted attention for their use as antimicrobial agents on foodstuff. In this regard, Chavez-López et al. (2012) proved that the use of terpenes from *O. vulgare* on fermented sausages are more effective against *A. fumigatus* without affecting the drying process, pH, water activity or color of the sausage. In contrast, Ashraf et al. (2017) reported that oregano terpenes had poor sensititivty against *A. niger*, *A. flavus*, *C. albicans*, *P. pinophilium*, and *S. cerevisiae* with around 15-40% of inhibition. Due to the possitive synergistic effect of oregano terpenes with terpenes from different matrixes, the blend of terpene-rich extracts from *O. vulgare*, *C. zeylanicum*, *D. carota*, and *S. aromaticum* was tested against six *Candida* strains (*C. albicans* DMS 1386, *C. glabatra* DSM 11226, *C. tropicalis* IP 2148.93, *C. albicans* F26, *C. albicans* f35, and *C. albicans* F78). Interestingly, the minimum inhibitory concentration ranged from 0.01% to 0.05% with minimal fungicidal concentrations from 0.02% to 0.05% (Brochot et al., 2017).

Terpenes from *O. vulgare* have also exhibited antifungal activity inhibiting mycelial growth against *Cladosporium cladosporoides*, *Alternaria alternata*, *Aspergillus fumigatus*, *Chaetomium globosum* and *Penicillium chrysogenum* with MIC values of 0.01% and 0.025% (Puškárová et al., 2017). *Origanum vulgare* terpenes are also active inhibitors of *Candida albicans*, *Cryptococcus neoformans*, *Trichophyton mentagrophytes*, *Trichophyton rubrum*, *Microsporum canis*, *Microsporum canis*, *Microsporum gypeum*, *Epidermophyton floccosum*, *Aspergillus fumigatus*, *Aspergillus fumigatus*, *Aspergillus flavus* (Vale-Silva et al., 2012). López et al. (2007) reported that essential oils from *O. vulgare*, inhibited foodborne fungi such as *C. albicans*, *Penicillium islandicum* and *Aspergillus flavus*.

Oregano terpenes have been evaluated as natural alternative inhibitors of fluconazole-susceptible *Candida*. In this subject, Pozzatti et al. (2008) showed that the terpenes from the oregano species *L. graveolens* and *O. vulgare* have effective antifungal potency against *C. albicans*, *C. dubliniensis*, *C. tropicalis*, *C. glabrata*, and *C. krusei*. Their main finding was that the susceptibilities of fluconazole-resistant *C. albicans*, *C. dubliniensis*, and *Candida* non-albicans were higher than those of the fluconazole-susceptible yeastes. Interestingly, carvacrol (56.8%) and *o*-cymene (32.2%) were the main constituents of *L. graveolens*, and carvacrol (92.6%) was the main component of the *O. vulgare* essential oil. Similarly, the antifungal activity of essential oils from *O. vulgare* and *L. graveolens* against *C. glabrata* has also been assessed by Soares et al. (2015); they used one group containing 30 fluconazole-susceptible *C. glabrata* isolates, and the second group contained fluconazole-resistant isolates. Antifungal activity less than or equal to 3200 $\mu g \cdot mL^{-1}$ was observed for both *O. vulgare* and *L. graveolens*. It is important to mention that terpenes from both species showed higher antifungal activity against the fluconazole-susceptible *C. glabrata* than against the fluconazole-resistant group.

Carvacrol is the terpene constituent to which antifungal activity has been attributed. It has been reported that the antifungal potency increases as carvacrol content increases. Similarly, y-terpinene content has been associate with the inhibition of filamentation (Vale-Silva et al., 2012). Even though the precise mechanism of action of terpenes is still controversial, alltogether the studies have suggested that terpenes penetrate into the microorganisms, interact with active sites of enzymes and/or interfere with the cellular metabolism. Another proposed mechanisms suggest less specific direct disruption of cell membranes and that this action is concentration-dependant (Bakkali et al., 2008; Vale-Silva et al., 2012). In this sense, the hydrophobicity alone of terpenes does not ensure their citotoxicity, for example, *p*-cymene, a precursor of carvacrol has a higher partition coefficient than carvacrol, however it is nontoxic (Rao et al., 2010). Instead, it has been suggested that the citotoxicity of terpenoids is given by a combination of factors, such as hydrophobicity and the presense of hydroxyl groups. This has been proposed by the lack of microbicidal effects of *p*-cymene and carvacrol methylesters. Which has led scientists to hipothetize that the delocalized electron system in phenolic terpenes, such as carvacrol, facilitates the dissociation of H^+ from the –OH group, dissipating pH and K^+ gradients across cell membranes. Carvacrol is also capable to depolarize bacterial cell membranes increasing membrane permeability (Rao et al., 2010; Ultee et al., 2002). Moreover, Ben Arfa et al. (2006) in agreement with Ultee et al. (2002), summarizes that "the presense of the hydroxyl group in phenolic terpenes allows compounds to act as proton exchanger, thereby reducing the gradient across the cytoplasmic membrane. The resulting collapse of the proton motrice force and depletion of the ATP pool lead eventually to cell death". In contrast, the hydroxyl moieties in carvacryl acetate and carvacrol methyl ether are not able to act as proton donors, which is reflected on its lower antimicrobial effect than carvacrol.

4.3. Antiviral Effect of Terpenes from Oregano

In addition to their antimicrobial, and anti-fungi properties, terpenes from oregano have also exhibited antiviral properties against some virus. Nonetheless, studies regarding antiviral properties of oregano terpenes are scarce. Even though there are few studies regarding antiviral properties of terpenoids from oregano, some authors have reported the antiviral activity of individual terpenes such as carvacrol, one of the main terpenes found in oregano. This could suggest the potential antiviral properties of all carvacrol-rich oregano species and varieties (Sánchez et al., 2015). However, further studies are needed since it has been shown that the interaction of all the constituents in terpene extracts from oregano can affect its biological properties (Leyva-López et al., 2017; Orhan et al., 2012).

Oregano terpenes have particularly been of interest in studies against the herpes simplex type-1 virus (HSV-1), which has caused concern among the medical community because the virus can remain latent in the nervous system and can be reactivated quite frequently. This is believed to be caused to the increased viral resistance caused by the prolonged use of acyvlovir and vidarabine, the main therapeutic agents. On this subject, Santoyo et al. (2014) showed that *O. vulgare* terpenes obtained by supercritical CO_2 were able to significantly inhibit the *in vitro* virus replication with IC50 values of 5.33 μg/mL. This action was attributed to carvacrol and thymol. On a similar note, Pilau et al. (2011) reported that *L. graveolens* and carvacrol are able to inhibit acyclovir-resistant herpes simplex virus type 1, acyclovir-sensitive HHV-1, and human respiratory syncytial virus. Similarly, Orhan et al. (2012) evaluated the antiviral properties against *Herpes simplex* type-1 (HSV-1) and parainfluenza type-3 (PI-3) of the terpene extracts obtained from the oregano species *O. onites*, *O. vulgare*, *O. munitiflorum* and *O. majorana*, and compared them to those of individual terpenes commonly found in those species. Interestingly, some individual terpenes were not as active against HSV type-1 as the extracts obtained from the tested oregano species; which suggests that the chemical composition of essential oils and the amount of individual terpenes are very important and related to each other in terms of biological activity.

Similarly, oregano terpenes have been tested against the yellow fever virus, which causes yellow fever (YF), a viral hemorrhagic fever. This virus is transmitted to humans through the bite of the *Aedes* or *Haemagogus* mosquitoes. YF causes at least 200,000 cases each year, including 30,000 deaths. And since, there is no antiviral drug against YF, the only way to prevent it is through vaccination. In this sense, Meneses et al. (2009) showed that terpenes from *Lippia alba*, *Lippia origanoides* and *O. vulgare* inhibited the yellow fever virus replication with a CC_{50}/MIC ratio of 22.9, 26.4, and 26.5, respectively. The authors suggested that the mode of action of the oregano species essential oils seems to be direct virus inactivation.

Another mode of action of oregano terpenes has been suggested by Gilling et al. (2014). The authors tested the antiviral activity of commercial oregano terpenes-rich extract against the nonenveloped murine norovirus; and reported that the terpenes acted directly upon the virus capsid and subsequently the RNA. Furthermore, under transmission electron microscopy, they revealed that the capsids were enlarged from < 35 nm in diameter to up to 75 nm following treatment with oregano terpenes. On the other hand, oregano terpenes did not inhibit the virus absorption to host cells. Table 4.1 summarizes the activity of terpenes from oregano against different bacteria, fungi and viruses.

4.4. Terpenes from Oregano as Antioxidant Agents

Oxidative stress results from an imbalance between the formation and detoxification of pro-oxidants. The presence of reactive species in excess of the available antioxidant capacity of animal cells, can modify several biologically cellular macromolecules such as lipids, proteins and DNA besides interfering with cell signaling pathways (Akbarian et al., 2016). Oxidative stress is associated with the development of numerous chronic degenerative diseases (Andrade et al., 2014).

Since the synthetic chemical compounds have been reported to have detrimental effects on human health, along with the increase in the demand of consumers for natural antioxidants, the research on plant-derived antioxidants becomes important. In this sense, terpenes, are plant-derived compounds, which have been found abundantly in oregano species, and their antioxidant potential has been investigated (Anastasaki et al., 2017).

The *in vitro* chemical assays including 2, 2-diphenyl-1-picrylhydrazyl (DPPH), 2,2'-azino-bis(3-ethylbenzothiazoline-6-sulphonic acid) (ABTS), Oxygen Radical Absorbance Capacity (ORAC), superoxide anion, reducing power (FRAP), hydroxyl-radical scavenging, among others, have been employed to investigate the antioxidant potential of extracts rich in terpenes from oregano, along with terpenes alone.

Extracts from *Origanum acutidens* (Hand-Mazz.) letswaart rich in carvacrol (61.69%) showed moderate radical scavenging activity (DPPH method) with a IC_{50} value of 387±5 μg•mL^{-1} increasing with concentrations (0.05-0.3 mg•mL^{-1}), the radical scavenging activity was compared to the synthetic antioxidants butylated hydroxytoluene (BHT) and butylated hydroxyanisole (BHA) (Öke Altuntaş & Demirtaş, 2017). Lower IC_{50} values indicate higher free radical scavenging activity.

Table 4.1. Antimicrobial properties of terpenes from oregano

Oregano species	Activity	Strain	Reference
L. alba	Antiviral	Yellow fever virus	(Meneses et al., 2009)
L. graveolens	Antimicrobial	*Bacillus subtilis*, *Staphylococcus aureus*, *E. coli*, *Salmonella typhi* and *Pseudomonas aeruginosa*	(Rivero-Cruz et al., 2011)
	Antimicrobial	*E. coli*, *Ps. aeruginosa* and *S. aureus*	(Arana-Sánchez et al., 2010)
	Antifungal	*Candida* spp. (*C. albicans*, *C. dubliniensis*, *C. tropicalis*, *C. glabrata*, and *C. krusei*)	(Pozzatti et al., 2008)
	Antiviral	Herpes simplex type-1	(Pilau et al., 2011)
L. origanoides	Antiviral	Yellow fever virus	(Meneses et al., 2009)
O. compactum	Antimicrobial	*S. aureus* MBLA, *Proteus mirabilis*	(Bouyahya et al., 2017)
	Antimicrobial	*E. coli*	(Van Haute et al., 2016)
O. heracleoticum	Antimicrobial	*E. coli*, *E. coli* O157:H7, *S. aureus*, *P. aureginosa* and *Lactobacillus plantarum*	(Emiroğlu et al., 2010)
O. majorana	Antiviral	Herpes simplex type-1, Parainfluenza type-3	(Orhan et al., 2012)
O. minutiflorum	Antiviral	Herpes simplex type-1, Parainfluenza type-3	(Orhan et al., 2012)
O. onites	Antimicrobial	*Salmonella typhimurium*	(Gündüz et al., 2010)
	Antiviral	Herpes simplex type-1, Parainfluenza type-3	(Orhan et al., 2012)
O. vulgare	Antimicrobial	*Salmonella enterica* and *Listeria monocytogenes*	(Mazzarrino et al., 2015)
	Antimicrobial	*Staphylococcus aureus*, *Streptococcus faecalis*, *Streptococcus pyogenes*, and *Listeria monocytogenes*, *Salmonella* spp. and *Shigella* spp.	(Grondona et al., 2014)

Table 4.1. (Continued)

Oregano species	Activity	Strain	Reference
	Antimicrobial	*S. aureus*, *B. cereus*, *E. fecalis*, *L. monocytogenes*, *E. coli*, *Yersinia enterocolitica*, *Salmonella cholerasuis* and *Pseudomonas aeruginosa*	(López et al., 2007)
	Antimicrobial	*S. aureus*, *B. cereus*, and *P. aeruginosa*	(S. M. B. Hashemi et al., 2017)
	Antimicrobial	*S. aureus* (ATCC 25923) and *E. coli* (ATCC 25922)	(Moraes-Lovison et al., 2017)
	Antimicrobial	*S. aureus*, *L. monocytogenes*, *E. coli* and *S. enteritidis*	(Benavides et al., 2012)
	Antimicrobial	*S. aureus* (ATCC 25923), *S. mutans* (ATCC 25175), *E. coli* (ATCC 25922), *S. enterica Enteritidis* (ATCC 13076), *K. pneumoniae* (ATCC 10031), *K. pneumoniae* (ATCC 700603), *S. enterica Typhimurium* UK-1, MRSA N315, MRSA 101, MRSA 107, *E. coli* ESBL 167, *E. coli* ESBL 169, *E. coli* ESBL 176, *E. coli* ESBL 192, *E. coli* KPC 131, *E. Coli* KPC 133, and *A. baumannii* CR 01	(Scandorieiro et al., 2016)
	Antifungal	*A. fumigatus*	(Chavez-López et al., 2012)
	Antifungal	*C. albicans* (DMS 1386, F26, f35, F78), *C. glabatra* DSM 11226, *C. tropicalis* IP 2148.93	(Ashraf et al., 2017)
	Antifungal	*C. cladosporoides*, *Alternaria alternata*, *A.fumigatus*, *Chaetomium globosum* and *Penicillium chrysogenum*	(Puškárová et al., 2017)

Oregano species	Activity	Strain	Reference
	Antifungal	*C. albicans*, *C. neoformans*, *T. mentagrophytes*, *Trichophyton rubrum*, *Microsporum canis*, *Microsporum canis*, *Microsporum gypeum*, *Epidermophyton floccosum*, *Aspergillus fumigatus*, *A. fumigatus*, *A. flavus*	(Vale-Silva et al., 2012)
	Antifungal	*Candida albicans*, *Penicillium islandicum* and *Aspergillus flavus*	(López et al., 2007)
	Antifungal	*Candida glabrata*	(Soares et al., 2015)
	Antifungal	*C. albicans*, *C. dubliniensis*, *C. tropicalis*, *C. glabrata*, and *C. krusei*	(Pozzatti et al., 2008)
	Antiviral	Herpes simplex type-1	(Santoyo et al., 2014)
	Antiviral	Herpes simplex type-1, Parainfluenza type-3	(Orhan et al., 2012)
	Antiviral	Yellow fever virus	(Meneses et al., 2009)
O. vulgare ssp. *gracile*	Antimicrobial	*E. coli* and *B. cereus*	(Jahed et al., 2017)
O. vulgare ssp. *hirtum*	Antimicrobial	*B. subtilis*, *C. albicans*, *E. faecalis*, *E. aerogenes*, *E. durans*, *E. faecium*, *E. coli*, *K. pneumoniae*, *L. monocytogenes*, *L. innocua*, *P. aeruginosa*, *P. fluorescence*, *S. infantis*, *S. kentucky*, *S. typhimurium*, *S. aureus* and *S. epidermidis*	(O. E. Ozkan et al., 2017)
	Antimicrobial	*Salmonella enteritidis*	(Govaris et al., 2010)
O. vulgare ssp. *viride*	Antimicrobial	*S. typhimurium*, *E. coli* O157:H7, *P. aeruginosa*, *S. aureus*, and *B. cereus*	(Seyed Mohammad Bagher Hashemi & Mousavi Khaneghah, 2017)

Table 4.1. (Continued)

Oregano species	Activity	Strain	Reference
P. longiflora	Antimicrobial	*Bacillus subtilis*, *Staphylococcus aureus*, *Escherichia coli*, *Salmonella typhi* and *Pseudomonas aeruginosa*	(Rivero-Cruz et al., 2011)
NR	Antimicrobial	*E. faecalis*, *S. aureus*, *S. mutans*, *Aeromonas sp*, *E. coli*, *K. pneumonae*, *S. choleraesius*, *Serratia sp, S. flexneri*	(Busatta et al., 2007)
NR	Antimicrobial	*S. typhimurium*, *P. aeruginosa*, *S. aureus*, *E. coli*, *Shewanella putrefaciens* and *Yersinia enterocolitica*	(Jouki et al., 2014)
NR	Antimicrobial	*Borrelia burgdorferri*	(Feng et al., 2017)

NR: Oregano species not specified.

A similar low radical scavenging activity against DPPH radical, was also reported by Albano et al. (2012), for extracts from *O. vulgare*, with a IC_{50} value of 156.3 ± 5.1 μg•mL^{-1}. This value was lower than the synthetic antioxidant BHT (IC_{50} = 13.0 ± 1.0 μg•mL^{-1}). The authors suggested that the antioxidant activity can be attributed to γ-terpinene and thymol present in *O. vulgare* extracts. A weak capacity to scavenge free radicals (DPPH) was reported for carvacrol-rich extracts from *O. dictamnus, Origanum libanoticum* and *Origanum microphyllum*, with values from 0.02 to 0.15 mmol of Trolox equivalents•L^{-1}, when compared to BHT (2.5 mmol of Trolox equivalents•L^{-1}) (Marrelli, Conforti, et al., 2015). In contrast, the same extracts showed better ferric reducing/antioxidant power (FRAP assay) with values from 0.50 to 0.57 mmol of Trolox equivalents•L^{-1}, however, these values were also lower than BHT (3.9 mmol of Trolox equivalents•L^{-1}). Spagnoletti et al. (2016) reported that thymol rich extracts from *O. hirtum* showed weak DPPH radical scavenging capacity with a IC_{50} value of 694 μg•mL^{-1}, when compared to the Trolox (5 μg•mL^{-1}) which was used as a reference standard, while in ABTS assay the extract exhibited strong antioxidant activity (IC_{50} = 32 μg•mL^{-1}), when compared to the same standard Trolox (IC_{50}= 42 μg•mL^{-1}). *O. majorana* L. extracts, which were rich in terpinen-4-ol (21%), showed significant antiradical activity with respect to the DPPH radical (IC_{50} = 341.0 μg•mL^{-1}) compared to thymol alone (IC_{50} = 560.0 μg•mL^{-1}). However, when compared to other synthetic antioxidant standards (rutin, ascorbic acid, BHT and BHA), the extract displayed negligible antiradical effect (Schmidt et al., 2008). In contrast, the same extract, showed powerful scavenge of $HO^{\bullet}$ (hydroxyl radical-scavenging activity assay), which reached the value of 73% (1 μg•mL^{-1}), and the effect was concentration dependent. Differences between the values in both assays may be attributed to the mechanism of the extract to neutralize the two radicals.

Another work, reported that, based on DPPH assay, *O. onites* L. extracts possess higher antioxidant potential (EC_{50} = 80 μg•mL^{-1}), than their major constituents thymol and carvacrol (EC_{50} of 163 and 248 μg•mL^{-1} respectively), while, the same extract, showed lower inhibition of the oxidation of the linoleic acid (40 %), than carvacrol (52.02 %) and thymol (57.11 %) (Özkan & Erdoğan, 2011).

Another work, showed that extracts from *O. vulgare* L. ssp. *vulgare,* possess high antioxidant activity with an IC_{50} value of 2.5 μg•mL^{-1} of radical scavenging activity against DPPH, at concentrations of 0.5, 5, 10 and 50 μg•mL^{-1}. This value was higher than the standard antioxidants Vitamin E and BHA. Likewise, results from ferric reducing antioxidant power (FRAP) confirmed the good antioxidant activity of essential oils (467.25 μmole Eq $FeSO_4.7H_2O$•mg^{-1} of the oil). The authors attribute the effect to the phenolic content, like thymol, that is a monoterpene phenol (Vazirian et al., 2015). DPPH assay indicated that carvacrol rich extract (65%) from *O. vulgare* ssp. *hirtum* had an IC_{50} value of 1.890±0.006 g·g^{-1} DPPH (Yaldiz et al., 2017).

The antioxidant potential of extracts obtained by hydrodistillation from *O. vulgare* ssp. *hirtum* was evalutated by the DPPH radical scavenging activity and the inhibition of linoleic acid peroxidation assays (Hodaj-Celiku et al., 2017). The authors reported carvacrol was the main constituent of the extracts (81 %). By comparison with other aromatic plants, *O. vulgare* ssp. *hirtum* showed the best scavenging ability (76.5 %), in contrast, the extract did not show any effect in the anti-lipid peroxidation activity.

A carvacrol rich extract (56.8 %) from *Origanum rotundifolium* Boiss exhibited different antioxidant capacities based on all six tests performed (Özbek et al., 2017). The extract (60 μg•mL^{-1}), showed a slight activity against DPPH with a value of 6.53%, and ferrous ion chelating activity (21.67%), which were lower than Trolox (45.92% and 59.83%, respectively), used as reference compound. The same extract also showed a strong ABTS radical cation decolorization activity (100%) and H_2O_2 scavenging activity (100%), which were higher than the reference compound (97.89% and 81.72% respectively). Slightly similar to Trolox (61.36%), the extracts showed a superoxide anion radical scavenging activity of 68.65%. Additionally, the inhibition of lipid peroxidation reached a value of 44.06%, which was a measure of the total antioxidant activity of the extract. The authors implied that the phenolic constituents of the extract may be responsible for the effects observed.

Another work investigated the genetic variability in antioxidant capacity from methanolic extracts (80%) of 42 accessions from different subspecies of *O. vulgare* L. (Yan et al., 2016). The samples exhibited significant variation in antioxidant activity tested by ORAC assay and the total phenolic content (TPC), measured by Folin-Ciocalteu method. ORAC assay measured the peroxyl radical scavenging capacity while the TPC measured the antioxidant's reducing capacity of the extracts. The differences between the highest and lowest value was 2-fold for ORAC and TPC. The predominant components of the essential oil from 42 oregano accessions were carvacrol (up to 66%), thymol (up to 66%), γ-terpinene (up to 11%) and *p*-cymene (up to 5.7%). The authors suggested that the variation may be explained in part by the variation on essential oil composition. Likewise, Daoudi-Merbah et al. (2016), reported that the location and the botanical characteristics may influence the antioxidant activity of six populations from *O. floribundum*. Although, carvacrol predominated in most populations, the sample with *p*-cymene as major component showed the highest antioxidant value. This sample, was able to scavenge DPPH and ABTS radicals with IC_{50} = 51.2 1± 0.3 and IC_{50} = 12.71 ± 0.37 μg•mL^{-1}. The authors showed that the antioxidant activity did not necessarily correlate with the content of carvacrol. Furthermore, they conclude that there must be a synergistic effect between the phenolic constituents (carvacrol and thymol) and *p*-cymene.

In contrast to other studies, δ- and γ-terpinene, and, oxygenated monoterpenes such as terpinen-4-ol and thymol were the main constituents of four different types of oregano (*Origanum x majoricum,* 'Mendocino'; *O. vulgare* ssp. *vulgare,* 'Compacto'; *O. vulgare* ssp. *hirtum clone* 'Cordobes'; and *O. vulgare* ssp. *hirtum* clone 'Criollo') from different regions. This extracts showed significant antioxidant activity based on four tests (Asensio et al., 2015). ABTS assay reached values ranging from 0.234 to 0.163 mM Trolox•mg^{-1} of essential oils. This value was highly correlated to thymol and terpinen-4-ol (R^2 = 0.42). The same four extracts also showed high ability to reduce iron (FRAP) with values ranging from

0.184 to 0.072 mM ascorbic acid/mg of essential oil, and, inhibition of bleaching (β-carotene bleaching assay), with values from by 75.3% to 89.2% (20 μg•mL^{-1} of essential oil), as well as high antioxidant potential in the ORAC value, with values from 1.708 to 1.024 Trolox Equivalents. Terpene rich extracts from of *O. onites* L. from different harvesting times scavenged DPPH and ABTS radicals with significant differences, with values from IC_{50} =116.74–132.93 μg•mL^{-1} and 23.54–31.02 mg ascorbic acid equivalent (AAE)•g^{-1} essential oil, respectively. The authors reported that highest values were observed in samples with higher amounts of carvacrol + thymol (94.63%) (G. Ozkan et al., 2010). In a like manner, Mechergui et al. (2016), reported the anti-radical activity (DPPH assay) of *O. grandulosum* collected at different harvesting years. The main compounds of the extracts were *p*-cymene, γ-terpinene, thymol and carvacrol, regardless of the year. Significant differences in the anti-radical activity were reported from one year to another in both populations tested. Nefza population showed values of IC_{50} from 59.2 to 226.019 mg•L^{-1} and for Krib population values of IC_{50} from 79.8 to 151.85 mg•L^{-1}. The differences may be attributed to the different composition of oregano extracts.

Terpene rich extracts *O. vulgare* L. showed similar antioxidant activities at different distillation times, based on the $ORAC_{oil}$. Extracts from the 20, 80, and 360 min had values of 60.8, 74.5, and 60.4 μmol Trolox equivalents•g^{-1}, respectively (Zheljazkov et al., 2012).

The antioxidant effect of nontoxic concentrations (95% cell viability) of *O. vulgare* extract was tested (Coccimiglio et al., 2016). Ethanolic extracts of oregano (0 to 2.93 μg•mL^{-1} of concentration) showed protective effect against hydrogen-peroxide induced cytotoxicity in A549 cells, in a concentration dependent manner. Additionally, pre-incubation of A549 cells with the isomeric monoterpene phenols, carvacrol or thymol, showed that the former monoterpene had a better protective effect.

Oregano species are rich in terpene compounds, mainly carvacrol and thymol, these compounds are predominantly components of essential oils of oregano, and, hence, the antioxidant potential of oregano essential oils has been attribute to terpenes compounds in several researches. The antioxidant effect of terpenes is the result of various possible mechanisms, such as, free-radical scavenging activity, transition-metal-chelating activity, and/or singlet-oxygen quenching capacity (Öke Altuntaş & Demirtaş, 2017).

4.5. Terpenes from Oregano and Their Anticancer Properties

Cancer is the second leading cause of death globally, and the number of new cases is expected to increase by around 70% over the next two decades (WHO, 2017). It is accepted that carcinogenesis is associated with oxidative stress, where reactive oxygen species are generated, therefore, antioxidant capacity of some phytochemicals like terpenes may be involved in cancer prevention by combating oxidative cell damages (Marrelli, Cristaldi, et al., 2015). Nowadays, there is a general call for new drugs with higher effectiveness, lower toxicity and the least negative environmental impact. Moreover, questions concerning the safety of synthetic drugs have increased the interest in the use of natural compounds from plant sources (Öke Altuntaş & Demirtaş, 2017).

Origanum vulgare extracts have been reported to possess anticancer activity, the main constituents of oregano are terpenes, and they possess different biological properties, such as antimutagenic, antiproliferative and antioxidant, which are responsible for its anticancer activity (Bhalla et al., 2013). The anticancer property of oregano species has been poorly studied (Pezzani et al., 2017), that is why the study of the potential application of oregano constituents to treat this disease is important.

The cytotoxicity of oregano extracts has been reported in different cell lines. For example, a recent work examined the cytotoxic effects of the essential oils from *O. acutidens* (Hand-Mazz.) letswaart. The authors used human colorectal adenocarcinoma (HT-29) and human cervical adenocarcinoma (HeLa) cell lines (using xCELLigence Real Time Cell Analyzer instrument), reporting significant cytotoxic effect against on HeLa cell lines, at concentrations of 50, 100 and 250 $\mu g \cdot mL^{-1}$ ($IC_{50} < 10$ $\mu g \cdot mL^{-1}$), and for HT-29 cell line at concentrations of 50 and 100 $\mu g \cdot mL^{-1}$.

They attribute the cytotoxic effect to carvacrol which is an important componcnt of oregano (Öke Altuntaş & Demirtaş, 2017).

Another work, evaluated the cytotoxic activity of carvacrol, thymol, *p*-cymene, 1-octacosanol alone and its combination. These compounds were isolated from *Origanum vulgare*. Using A549 human lung adenocarcinoma epithelial cells, the authors reported a decrease in cell viability in a concentration dependent manner, and thymol as the most cytotoxic agent. Furthermore, the authors suggested that the presence of the terpenes carvacrol and thymol may be the responsible for the effects (Coccimiglio et al., 2016).

HepG2 cell line has been also evaluated (Özkan & Erdoğan, 2011). The authors reported that the cytotoxicity of carvacrol and thymol isolated from *O. onites* L. was of IC_{50}= 53.09 and 60 μg•mL^{-1} respectively. By comparison with essential oil extract, the isolated compounds were more cytotoxic. Also, the antiproliferative effects on HepG2 cells, were dose-dependent. The authors conclude that the essential oil's components had anticancer effects. Antiproliferative effect of essential oils from *O. dictamnus, Origanum libanoticum* and *Origanum* microphyllum against the two human cancer cell lines LoVo (colon carcinoma) and HepG2 (hepatocarcinoma) was also investigated (Marrelli, Conforti, et al., 2015). The authors reported that at 100 μg•mL^{-1}, essential oils extracted from *O. dictamus* showed the best biological activity with an inhibition on LoVo cell line of 58.39% after 24 h. Furthermore, the IC_{50} value for this essential oil was 84.76 μg•mL^{-1} after 24 h and 72.26 μg•mL^{-1} after 48 h of treatment. Additionally, it was reported an antiproliferative activity for HepG2 cell line of 49.83% y 48.50% for a concentration of 100 μg•mL^{-1} of *O. dictamus* and *O. libanoticum* essential oil's respectively.

4.6. Anti-Inflammatory Activity of Terpenes Found in Oregano

Inflammation is the response of the organism to tissue damage, infections and chemical or physical agents (Medzhitov, 2008). During this process the synthesis of inflammatory mediators is initiated. Some of the most important mediators in inflammation are enzymes, prostaglandins, cytokines, nitric oxide (NO) and reactive oxygen species (ROS), amongst others (V. Kumar et al., 2013). If the process of inflammation is not regulated adequately, the mediators are overproduced which might lead to pathologic processes such as arthritis, atherosclerosis and cancer (Hansson 2005; Trinchieri, 2012). Therefore, inhibition of the mediators is a very important aim to treat inflammation-related diseases.

There is evidence that mention that terpenes have anti-inflammatory potential. For instance, Leyva-López et al. (2016) established that terpenes, such as thymol and carvacrol acetate, obtained from the three Mexican oregano species, *L. graveolens*, *L. palmeri* and *H. patens* reduced significantly the levels of ROS and NO produced by RAW 264.7 macrophage cells stimulated with lipopolysaccharide (LPS). Besides, EOs of *O. majorana* (10 μg/mL), containing the terpenes *cis*-sabinene hydrate and terpinen-4-acetate, were able to reduce the levels of tumor necrosis factor-alpha (TNF-α), interleukin-1β (IL-1β) and IL-6 in LPS-activated THP-1 human macrophage cells (Arranz et al., 2015). Recently, X. Han and Parker (2017) demonstrated that essential oils from *O. vulgare,* where the main constituent was the monoterpenoid carvacrol (78%), inhibited the production of the intracellular cell adhesion molecule-1 (ICAM-1), vascular cell adhesion molecule-1 (VCAM-1) and the monocyte chemoattractan protein-1 (MCP-1) on activated-primary human neonatal fibroblasts, which are known as inflammation-related biomarkers. These results propose that terpenes from oregano possess anti-inflammatory potential. Additionally, Lima et al. (2013) established that carvacrol exerts anti-inflammatory action on a common inflammation model. Carvacrol was administrated to mice with paw edema at 50 and 100 $mg{\bullet}kg^{-1}$. In this study it was found that the levels of IL-1β and prostaglandin E_2 were reduced and that the decrease on the mRNA expression of IL-1β and cyclooxygenase-2 might be responsible for the effects mentioned. On the other hand, the level of the IL-10 in the inflamed paw was upgraded by carvacrol. Therefore, the anti-inflammatory action of carvacrol is due to the reduction of pro-inflammatory mediators but also to the increasing of anti-inflammatory cytokines (IL-10) (Lima et al., 2013). Carvacrol has also showed the ability to prevent obesity in mice by modulating expression of genes involved in inflammation (Cho et al., 2012) and to attenuate induced liver injury and ulcer in rats (Rajan et al., 2015; Silva et al., 2012). Other terpenes such as *p*-cymene (de Souza Siqueira Quintans et al., 2013) and β-Caryophyllene (Abbas et al., 2013; Alberti et al., 2017; Klauke et al., 2014; Park et al., 2011) have also demonstrated anti-inflammatory activity.

The researches cited above indicate that oregano species could be employed to formulate anti-inflammatory agents for both, the prevention and the treatment of inflammation-related diseases. Nevertheless clinical studies are needed before terpenes can be used as an alternative to treat inflammation.

Conclusion

Oregano species are plants with a vast diversity in terms of composition and biological activities. In this chapter it has been established that this herb might be used for extraction of a great variety of terpenes with use in food, cosmetic and pharmaceutical industries. Nevertheless, it has to be noted that several factors might affect the main constituents of oregano and so their potential to be used natural agents to treat diseases. Also, there is the need to evaluate the effects of the use of terpenes from oregano in humans by clinical studies. Therefore, the correct analysis of the composition and the understanding of the biological activities of oregano might represent a very good opportunity to continue investigating the potential health benefits of oregano species and to take advantage into the development of a wide variety of new bioactive or nutraceutical products.

References

Abbas, M. A., Taha, M. O., Zihlif, M. A. & Disi, A. M. (2013). β-Caryophyllene causes regression of endometrial implants in a rat model of endometriosis without affecting fertility. *European Journal of Pharmacology*, *702*(1–3), 12-19. doi: http://doi.org/10.1016/j.ejphar.2013.01.011.

Acevedo, D., Navarro, M. & Monroy, L. (2013). Composición Química del aceite escencial de hojas de orégano (*Origanum vulgare*). *Información tecnológica*, *24*(4), 43-48. [Chemical composition of essential oil from oregano leaves. *Technological Information*, *24*(4), 43-48].

Adame-Gallegos, J. R., Andrade-Ochoa, S. & Nevarez-Moorillon, G. V. (2016). Potential use of Mexican oregano essential oil against parasite, fungal and bacterial pathogens. *Journal of Essential Oil-Bearing Plants*, *19*(3), 553-567. doi: 10.1080/0972060X. 2015.1116413.

Akbarian, A., Michiels, J., Degroote, J., Majdeddin, M., Golian, A. & De Smet, S. (2016). Association between heat stress and oxidative stress in poultry; mitochondrial dysfunction and dietary interventions with phytochemicals. *Journal of Animal Science and Biotechnology*, *7*, 37. doi: 10.1186/s40104-016-0097-5.

Al Hafi, M., El Beyrouthy, M., Ouaini, N., Stien, D., Rutledge, D. & Chaillou, S. (2016). Chemical composition and antimicrobial activity of *Origanum libanoticum*, *Origanum ehrenbergii*, and *Origanum syriacum* growing wild in Lebanon. *Chemistry & Biodiversity*, *13*(5), 555-560. doi: 10.1002/cbdv.201500178.

Albano, S. M., Lima, A. S., Pedro, M. G. M. L. G., Barroso, J. G. & Figueiredo, A. C. (2012). Antioxidant, anti-5-lipoxygenase and antiacetylcholinesterase activities of essential oils and decoction waters of some aromatic plants. *Records of Natural Products*, *6*(1), 35.

Alberti, T., Barbosa, W., Vieira, J., Raposo, N. & Dutra, R. (2017). (−)-β-Caryophyllene, a CB2 Receptor-Selective phytocannabinoid, suppresses motor paralysis and neuroinflammation in a murine model of multiple sclerosis. *International Journal of Molecular Sciences*, *18*(4), 691.

Alor-Chávez, M., Estrada-Andrade, L., Aparicio-Trápala, M. & López-Hernández, E. (2017). Análisis cualitativo de metabolitos por cromatografía en capa fina en plantas. *Aportaciones a las Ciencias*

Alimentarias, 188. [Qualitative analysis of plant metabolites by use of thin layer chromatography. *Food Sciences Inputs*. 188].

Amadio, C., Medina, R., Dediol, C., Zimmermann, M. É. & Miralles, S. (2011). Aceite esencial de orégano: un potencial aditivo alimentario. *Revista de la Facultad de Ciencias Agrarias. Universidad Nacional de Cuyo*, *43*(1), 237-245. [Oregano essential oil: a potential food additive. *Journal of the Agricultural Sciences School. Cuyo National University*. *43*(1), 237-245]

Anastasaki, E., Zoumpopoulou, G., Astraka, K., Kampoli, E., Skoumpi, G., Papadimitriou, K., . . . Polissiou, M. (2017). Phytochemical analysis and evaluation of the antioxidant and antimicrobial properties of selected herbs cultivated in Greece. *Industrial Crops and Products*, *108* (Supplement C), 616-628. doi: https://doi.org/10.1016/j.indcrop.2017. 06.066.

Andrade, B. F. M. T., Braga, C. P., dos Santos, K. C., Barbosa, L. N., Rall, V. L. M., Sforcin, J. M., . . . Fernandes Júnior, A. (2014). Effect of inhaling *Cymbopogon martinii* essential oil and geraniol on serum biochemistry parameters and oxidative stress in rats. *Biochemistry Research International*, *2014*.

Arana-Sánchez, A., Estarrón-Espinosa, M., Obledo-Vázquez, E. N., Padilla-Camberos, E., Silva-Vázquez, R. & Lugo-Cervantes, E. (2010). Antimicrobial and antioxidant activities of Mexican oregano essential oils (*Lippia graveolens* H. B. K.) with different composition when microencapsulated in β-cyclodextrin. *Letters in Applied Microbiology*, *50*(6), 585-590. doi: 10.1111/j.1472-765X.2010.02837.x.

Arranz, E., Jaime, L., López de las Hazas, M. C., Reglero, G. & Santoyo, S. (2015). Supercritical fluid extraction as an alternative process to obtain essential oils with anti-inflammatory properties from marjoram and sweet basil. *Industrial Crops and Products*, *67*, 121-129. doi: http://doi.org/10.1016/j.indcrop.2015.01.012.

Asensio, C. M., Grosso, N. R. & Juliani, H. R. (2015). Quality characters, chemical composition and biological activities of oregano (*Origanum* spp.) Essential oils from Central and Southern Argentina. *Industrial Crops and Products*, *63*, 203-213.

Ashraf, S. A., Al-Shammari, E., Hussain, T., Tajuddin, S. & Panda, B. P. (2017). In-vitro antimicrobial activity and identification of bioactive components using GC–MS of commercially available essential oils in Saudi Arabia. *Journal of Food Science and Technology*, *54*(12), 3948-3958. doi: 10.1007/s13197-017-2859-2.

Azizi, A., Yan, F. & Honermeier, B. (2009). Herbage yield, essential oil content and composition of three oregano (*Origanum vulgare* L.) populations as affected by soil moisture regimes and nitrogen supply. *Industrial Crops and Products*, *29*(2), 554-561. doi: https://doi.org/10.1016/j.indcrop.2008.11.001.

Bakkali, F., Averbeck, S., Averbeck, D. & Idaomar, M. (2008). Biological effects of essential oils – A review. *Food and Chemical Toxicology*, *46*(2), 446-475. doi: 10.1016/j.fct.2007.09.106.

Baranauskaitė, J., Jakštas, V., Ivanauskas, L., Kopustinskienė, D. M., Drakšienė, G., Masteikova, R. & Bernatonienė, J. (2016). Optimization of carvacrol, rosmarinic, oleanolic and ursolic acid extraction from oregano herbs (*Origanum onites* L., *Origanum vulgare* spp. hirtum and *Origanum vulgare* L.). *Natural Product Research*, *30*(6), 672-674. doi: 10.1080/14786419.2015.1038998.

Baranauskienė, R., Venskutonis, P. R., Dambrauskienė, E. & Viškelis, P. (2013). Harvesting time influences the yield and oil composition of *Origanum vulgare* L. ssp. vulgare and ssp. hirtum. *Industrial Crops and Products*, *49*, 43-51. doi: https://doi.org/10.1016/ j.indcrop. 2013.04.024.

Baser, K. H. C. & Buchbauer, G. (2015). *Handbook of essential oils: science, technology, and applications*: CRC Press.

Bayramoglu, B., Sahin, S. & Sumnu, G. (2008). Solvent-free microwave extraction of essential oil from oregano. *Journal of Food Engineering*, *88*(4), 535-540.

Begnini, K. R., Nedel, F., Lund, R. G., Carvalho, P. H. d. A., Rodrigues, M. R. A., Beira, F. T. A. & Del-Pino, F. A. B. (2014). Composition and antiproliferative effect of essential oil of *Origanum vulgare* against tumor cell lines. *Journal of Medicinal Food*, *17*(10), 1129-1133.

Ben Arfa, A., Combes, S., Preziosi-Belloy, L., Gontard, N. & Chalier, P. (2006). Antimicrobial activity of carvacrol related to its chemical structure. *Letters in Applied Microbiology*, *43*(2), 149-154. doi: 10.1111/j.1472-765X.2006.01938.x.

Benavides, S., Villalobos-Carvajal, R. & Reyes, J. E. (2012). Physical, mechanical and antibacterial properties of alginate film: Effect of the crosslinking degree and oregano essential oil concentration. *Journal of Food Engineering*, *110*(2), 232-239. doi: 10.1016/j.jfoodeng. 2011.05 .023.

Bertelli, D., Plessi, M. & Miglietta, F. (2003). Effect of microwaves on volatile compounds in origanum. *LWT-Food Science and Technology*, *36*(6), 555-560.

Betancourt, L., Phandanauvong, V., Patiño, R., Ariza-Nieto, C. & Afanador-Téllez, G. (2012). Composition and bactericidal activity against beneficial and pathogenic bacteria of oregano essential oils from four chemotypes of *Origanum and Lippia* genus. *Revista de la Facultad de Medicina Veterinaria y de Zootecnia*, *59*, 21-31.

Bhalla, Y., Gupta, V. K. & Jaitak, V. (2013). Anticancer activity of essential oils: a review. *Journal of the Science of Food and Agriculture*, *93*(15), 3643-3653.

Bonfanti, C., Iannì, R., Mazzaglia, A., Lanza, C. M., Napoli, E. M. & Ruberto, G. (2012). Emerging cultivation of oregano in Sicily: Sensory evaluation of plants and chemical composition of essential oils. *Industrial Crops and Products*, *35*(1), 160-165.

Bouyahya, A., Dakka, N., Talbaoui, A., Et-Touys, A., El-Boury, H., Abrini, J. & Bakri, Y. (2017). Correlation between phenological changes, chemical composition and biological activities of the essential oil from Moroccan endemic Oregano (*Origanum compactum* Benth). *Industrial Crops and Products*, *108* (Supplement C), 729-737. doi: 10.1016/j.indcrop.2017.07.033.

Bower, A. M., Real Hernandez, L. M., Berhow, M. A. & de Mejia, E. G. (2014). Bioactive Compounds from culinary herbs inhibit a molecular target for type 2 diabetes management, Dipeptidyl Peptidase IV.

Journal of Agricultural and Food Chemistry, *62*(26), 6147-6158. doi: 10.1021/jf500639f.

Bozin, B., Mimica-Dukic, N., Simin, N. & Anackov, G. (2006). Characterization of the volatile composition of essential oils of some Lamiaceae spices and the antimicrobial and antioxidant activities of the entire oils. *Journal of Agricultural and Food Chemistry*, *54*(5), 1822-1828.

Brochot, A., Guilbot, A., Haddioui, L. & Roques, C. (2017). Antibacterial, antifungal, and antiviral effects of three essential oil blends. *Microbiology Open*, *6*(4), e00459-n/a. doi: 10.1002/mbo3.459.

Bunghez, F., Rotar, M.-A., Pop, R. M., Romanciuc, F., Csernatoni, F., Fetea, F., . . . Socaciu, C. (2015). Comparative phenolic fingerprint and LC-ESI+ QTOF-MS composition of oregano and rosemary hydrophilic extracts in relation to their antibacterial effect. *Bulletin UASVM Food Science and Technology*, *72*, 1.

Burt, S. (2004). Essential oils: their antibacterial properties and potential applications in foods—a review. *International Journal of Food Microbiology*, *94*(3), 223-253. doi: 10.1016/j.ijfoodmicro.2004.03.022.

Busatta, C., Mossi, A. J., Rodrigues, M. R. A., Cansian, R. L. & de Oliveira, J. V. (2007). Evaluation of *Origanum vulgare* essential oil as antimicrobial agent in sausage. *Brazilian Journal of Microbiology*, *38*(4), 610-616. doi: 10.1590/s1517-83822007000400006.

Calvo-Irabién, L. M., Parra-Tabla, V., Acosta-Arriola, V., Escalante-Erosa, F., Díaz-Vera, L., Dzib, G. R. & Peña-Rodríguez, L. M. (2014). Phytochemical diversity of the essential oils of Mexican oregano (*Lippia graveolens* Kunth) populations along an edapho-climatic gradient. *Chemistry & Biodiversity*, *11*(7), 1010-1021.

Capuzzo, A., Maffei, M. E. & Occhipinti, A. (2013). Supercritical fluid extraction of plant flavors and fragrances. *Molecules*, *18*(6), 7194-7238.

Castilho, P. C., Savluchinske-Feio, S., Weinhold, T. S. & Gouveia, S. C. (2012). Evaluation of the antimicrobial and antioxidant activities of essential oils, extracts and their main components from oregano from Madeira Island, Portugal. *Food Control*, *23*(2), 552-558.

Clevenger, J. (1928). Apparatus for the determination of volatile oil. *Journal of Pharmaceutical Sciences, 17*(4), 345-349.

Coccimiglio, J., Alipour, M., Jiang, Z.-H., Gottardo, C. & Suntres, Z. (2016). Antioxidant, antibacterial, and cytotoxic activities of the ethanolic *Origanum vulgare* extract and its major constituents. *Oxidative Medicine and Cellular Longevity, 2016.*

Crocoll, C., Asbach, J., Novak, J., Gershenzon, J. & Degenhardt, J. (2010). Terpene synthases of oregano (*Origanum vulgare* L.) and their roles in the pathway and regulation of terpene biosynthesis. *Plant Molecular Biology, 73*(6), 587-603.

Chavez-López, C., Martin-Sanchez, A. M., Fuentes-Zaragosa, E., Viuda-Martos, M., Fernández-López, J., Sendra, E., . . . Pérez-Alvarez, J. A. (2012). Role of oregano (*Origanum vulgare*) essential oil as a surface fungus inhibitor on fermented sausages: evaluation of its effect on microbial and physicochemical characteristics. *Journal of Food Protection, 75*(1), 104-111. doi: 10.4315/0362-028x.jfp-11-184.

Chinou, I. (2010). Herbak drugs and the role of chromatographic methods in their analysis. In M. Waksmundzka-Hajnos & J. Sherma (Eds.), *High Performance Liquid Chromatography in Phytochemical Analysis*: CRC Press.

Chishti, S. (2016). Quantitative analysis of the essential oil of *Origanum vulgare* (L) growing in the Kashmir valley. *International Journal of Applied Research, 2*(2), 552-557.

Cho, S., Choi, Y., Park, S. & Park, T. (2012). Carvacrol prevents diet-induced obesity by modulating gene expressions involved in adipogenesis and inflammation in mice fed with high-fat diet. *The Journal of Nutritional Biochemistry, 23*(2), 192-201. doi: http://doi.org/10.1016/j.jnutbio.2010.11.016.

Daoudi-Merbah, F., Hazzit, M. & Dahmani-Megrerouche, M. (2016). Influence of morphological variability and habitat on the chemical composition of essential oils of an Algerian endemic *Origanum* species (*Origanum floribundum,* Munby). *Chemistry & Biodiversity, 13*(8), 1088-1094.

De Falco, E., Mancini, E., Roscigno, G., Mignola, E., Taglialatela-Scafati, O. & Senatore, F. (2013). Chemical composition and biological activity of essential oils of *Origanum vulgare* L. subsp. *vulgare* L. under different growth conditions. *Molecules*, *18*(12), 14948-14960.

de Souza Siqueira Quintans, J., Menezes, P. P., Santos, M. R. V., Bonjardim, L. R., Almeida, J. R. G. S., Gelain, D. P., . . . Quintans-Júnior, L. J. (2013). Improvement of p-cymene antinociceptive and anti-inflammatory effects by inclusion in β-cyclodextrin. *Phytomedicine*, *20*(5), 436-440. doi: http://doi.org/10.1016/j.phymed.2012.12.009.

Dugo, P., Mondello, L. & Dugo, G. (2000). Essential oils thin-Layer (Planar) Chromatography In: *Encyclopedia of Separation Science*. Academic Press. Pp 4502. DOI: 10.1016/B0-12-226770-2/02711-3.

Emiroğlu, Z. K., Yemiş, G. P., Coşkun, B. K. & Candoğan, K. (2010). Antimicrobial activity of soy edible films incorporated with thyme and oregano essential oils on fresh ground beef patties. *Meat Science*, *86*(2), 283-288. doi: 10.1016/j.meatsci.2010.04.016.

Feng, J., Zhang, S., Shi, W., Zubcevik, N., Miklossy, J. & Zhang, Y. (2017). Selective essential oils from spice or culinary herbs have high activity against stationary phase and biofilm *Borrelia burgdorferi*. *Frontiers in Medicine*, *4*(169). doi: 10.3389/fmed.2017.00169.

Figueredo, G., Musa Özcan, M., Chalchat, J. C., Bagci, Y. & Chalard, P. (2012). Chemical composition of essential oil of *Hyssopus officinalis* L. and *Origanum acutidens*. *Journal of Essential Oil Bearing Plants*, *15*(2), 300-306. doi: 10.1080/0972060X.2012.10644051.

Flanagan, R. J., Taylor, A., Watson, I. D. & Whelpton, R. (2008). Thin-layer chromatography. *Fundamentals of Analytical Toxicology* (pp. 131-143): John Wiley & Sons, Ltd.

Fornari, T., Ruiz-Rodriguez, A., Vicente, G., Vázquez, E., García-Risco, M. R. & Reglero, G. (2012). Kinetic study of the supercritical CO_2 extraction of different plants from Lamiaceae family. *The Journal of Supercritical Fluids*, *64*, 1-8.

Fornari, T., Vicente, G., Vázquez, E., García-Risco, M. R. & Reglero, G. (2012). Isolation of essential oil from different plants and herbs by supercritical fluid extraction. *Journal of Chromatography A*, *1250*, 34-48.

Gil-Chávez, J. G., Heredia, J. B. & González-Aguilar, G. A. (2012). Fluidos supercriticos: una alternativa útil y eficaz para la extracción y aplicación de antioxidantes en la industria de alimentos. *In: Antioxidantes en Alimentos y Salud* (pp. 559-584). AM Editores. ISBN978-607-437-207-6. [Supercritical fluids: an efficient and useful alternative to extract and use of antioxidants for food industry. In: *Antioxidants in Food and Heatlh* (pp. 559-584). AM Editors. ISBN978-607-437-207-6].

Gil-Chávez, G., Villa, J. A., Fernando Ayala-Zavala, J., Basilio Heredia, J., Sepulveda, D., Yahia, E. M. & González-Aguilar, G. A. (2013). Technologies for extraction and production of bioactive compounds to be used as nutraceuticals and food ingredients: an overview. *Comprehensive Reviews in Food Science and Food Safety*, *12*(1), 5-23.

Gilling, D. H., Kitajima, M., Torrey, J. R. & Bright, K. R. (2014). Antiviral efficacy and mechanisms of action of oregano essential oil and its primary component carvacrol against murine norovirus. *Journal of Applied Microbiology*, *116*(5), 1149-1163. doi: 10.1111/jam.12453.

Govaris, A., Solomakos, N., Pexara, A. & Chatzopoulou, P. S. (2010). The antimicrobial effect of oregano essential oil, nisin and their combination against Salmonella Enteritidis in minced sheep meat during refrigerated storage. *International Journal of Food Microbiology*, *137*(2), 175-180. doi: 10.1016/j.ijfoodmicro. 2009.12.017.

Grández, V. & Einstein, A. (2017). *Comparación de tres métodos en la extracción de aceite esencial de orégano silvestre (Lippia ssp.).* Universidad Nacional Toribio Rodríguez de Mendoza de Amazonas. Tesis. Pp. 34 [*Comparison of three methods for extraction of wild oregano (Lippia spp) essential oils*. Toribio Rodriguez de Mendoza de Amazonas National University. Thesis. Pp 34].

Grevsen, K., Frett, xe, C., X. & Christensen, L. P. (2009). Content and composition of volatile terpenes, flavonoids and phenolic acids in Greek oregano (*Origanum vulgare* L. ssp. hirtum) at different development stages during cultivation in cool temperate climate. *European Journal of Horticultural Science, 74*(5), 193-203.

Grondona, E., Gatti, G., López, A. G., Sánchez, L. R., Rivero, V., Pessah, O., . . . Ponce, A. A. (2014). Bio-efficacy of the essential oil of oregano (*Origanum vulgare* Lamiaceae. Ssp. Hirtum). *Plant Foods for Human Nutrition, 69*(4), 351-357. doi: 10.1007/s11130-014-0441-x.

Gündüz, G. T., Gönül, Ş. A. & Karapınar, M. (2010). Efficacy of oregano oil in the inactivation of *Salmonella typhimurium* on lettuce. *Food Control, 21*(4), 513-517. doi: 10.1016/j.foodcont.2009.07.016.

Hajnos, M. Ł. (2010). HPLC Analysis of Diterpenes. In M. Waksmundzka-Hajnos & J. Sherma (Eds.), *High Performance Liquid Chromatography in Phytochemical Analysis*: CRC Press.

Han, F., Ma, G.-q., Yang, M., Yan, L., Xiong, W., Shu, J.-c., . . . Xu, H.-l. (2017). Chemical composition and antioxidant activities of essential oils from different parts of the oregano. *Journal of Zhejiang University. Science. B, 18*(1), 79.

Han, X. & Parker, T. L. (2017). Anti-inflammatory, tissue remodeling, immunomodulatory, and anticancer activities of oregano (*Origanum vulgare*) essential oil in a human skin disease model. *Biochimie Open, 4*, 73-77. doi: http://doi.org/10.1016/j.biopen.2017.02.005.

Hansson, G. K. (2005). Inflammation, atherosclerosis, and coronary artery disease. *New England Journal of Medicine, 352*(16), 1685-1695. doi: doi:10.1056/NEJMra043430.

Hashemi, S. M. B. & Mousavi Khaneghah, A. (2017). Characterization of novel basil-seed gum active edible films and coatings containing oregano essential oil. *Progress in Organic Coatings, 110* (Supplement C), 35-41. doi: 10.1016/j.porgcoat.2017.04.041.

Hashemi, S. M. B., Nikmaram, N., Esteghlal, S., Khaneghah, A. M., Niakousari, M., Barba, F. J., . . . Koubaa, M. (2017). Efficiency of Ohmic assisted hydrodistillation for the extraction of essential oil from oregano (*Origanum vulgare* subsp viride) spices. *Innovative Food Science & Emerging Technologies, 41*, 172-178. doi: 10.1016/ j.ifset.2017.03.003.

Hernández, H., Fraňková, A., Sýkora, T., Klouček, P., Kouřimská, L., Kučerová, I. & Banout, J. (2017). Thc cffcct of orcgano csscntial oil on microbial load and sensory attributes of dried meat. *Journal of the Science of Food and Agriculture*, *97*(1), 82-87. doi: 10.1002/jsfa.7685.

Hodaj-Celiku, E., Tsiftsoglou, O., Shuka, L., Abazi, S., Hadjipavlou-Litina, D. & Lazari, D. (2017). Antioxidant activity and chemical composition of essential oils of some aromatic and medicinal plants from Albania. *Natural Product Communications*, *12*(5), 785-790.

Hossain, M. B., Patras, A., Barry-Ryan, C., Martin-Diana, A. B. & Brunton, N. P. (2011). Application of principal component and hierarchical cluster analysis to classify different spices based on in vitro antioxidant activity and individual polyphenolic antioxidant compounds. *Journal of Functional Foods*, *3*(3), 179-189. doi: https://doi.org/10.1016/j.jff.2011.03.010.

Hossain, M. B., Rai, D. K., Brunton, N. P., Martin-Diana, A. B. & Barry-Ryan, C. (2010). Characterization of phenolic composition in Lamiaceae spices by LC-ESI-MS/MS. *Journal of Agricultural and Food Chemistry*, *58*(19), 10576-10581.

Ivanovic, J., Zizovic, I., Ristic, M., Stamenic, M. & Skala, D. (2011). The analysis of simultaneous clove/oregano and clove/thyme supercritical extraction. *The Journal of Supercritical Fluids*, *55*(3), 983-991.

Jahed, E., Khaledabad, M. A., Bari, M. R. & Almasi, H. (2017). Effect of cellulose and lignocellulose nanofibers on the properties of *Origanum vulgare* ssp. gracile essential oil-loaded chitosan films. *Reactive and Functional Polymers*, *117* (Supplement C), 70-80. doi: 10.1016/ j.reactfunctpolym.2017.06.008.

Jayasena, D. D. & Jo, C. (2013). Essential oils as potential antimicrobial agents in meat and meat products: A review. *Trends in Food Science & Technology*, *34*(2), 96-108. doi: 10.1016/j.tifs.2013.09.002.

Jiang, Z., Kempinski, C. & Chappell, J. (2016). Extraction and analysis of terpenes/terpenoids. *Current protocols in plant biology*, 345-358.

Jouki, M., Yazdi, F. T., Mortazavi, S. A. & Koocheki, A. (2014). Quince seed mucilage films incorporated with oregano essential oil: Physical, thermal, barrier, antioxidant and antibacterial properties. *Food*

Hydrocolloids, *36* (Supplement C), 9-19. doi: 10.1016/ j.foodhyd. 2013.08.030.

Kintzios, S. E. (2003). *Oregano: the genera Origanum and Lippia*: CRC press. Pp 267.

Kintzios, S. E. (2012). 21 - Oregano A2 - Peter, K.V *Handbook of Herbs and Spices (Second edition)* (pp. 417-436): Woodhead Publishing.

Kirby, J. & Keasling, J. D. (2009). Biosynthesis of plant isoprenoids: perspectives for microbial engineering. *Annual Review of Plant Biology*, *60*(1), 335-355. doi: 10.1146/annurev.arplant.043008.091955.

Klauke, A. L., Racz, I., Pradier, B., Markert, A., Zimmer, A. M., Gertsch, J. & Zimmer, A. (2014). The cannabinoid CB2 receptor-selective phytocannabinoid beta-caryophyllene exerts analgesic effects in mouse models of inflammatory and neuropathic pain. *European Neuropsychopharmacology*, *24*(4), 608-620. doi: http://doi.org/ 10.1016/j.euroneuro.2013.10.008.

Koch, A., Basar, S. & Richter, R. (2010). HPLC of Mono- and Sesquiterpenes. In M. Waksmundzka-Hajnos & J. Sherma (Eds.), *High Performance Liquid Chromatography in Phytochemical Analysis*. CRC Press. Pp 996.

Kogiannou, D. A. A., Kalogeropoulos, N., Kefalas, P., Polissiou, M. G. & Kaliora, A. C. (2013). Herbal infusions; their phenolic profile, antioxidant and anti-inflammatory effects in HT29 and PC3 cells. *Food and Chemical Toxicology*, *61* (Supplement C), 152-159. doi: https://doi.org/10.1016/j.fct.2013.05.027.

Kumar, R. & Tripathi, Y. (2011). Getting fragrance from plants. *Training Manual on Extraction Technology of Natural Dyes & Aroma Therapy and Cultivation Value Addition of Medicinal Plants*, 77-102.

Kumar, V., Abbas, A. K. & Aster, J. C. (2013). *Robbins Basic Pathology*: Elsevier Health Sciences.

Laferriere, J. E., Weber, C. W. & Kohlhepp, E. A. (1991). Use and nutritional composition of some traditional Mountain Pima plant foods. *Journal of Ethnobiology*, *11*(1), 93-114.

Leyva-López, N., Gutiérrez-Grijalva, E., Vazquez-Olivo, G. & Heredia, J. (2017). Essential Oils of Oregano: Biological Activity beyond Their Antimicrobial Properties. *Molecules*, *22*(6), 989.

Leyva-López, N., Nair, V., Bang, W. Y., Cisneros-Zevallos, L. & Heredia, J. B. (2016). Protective role of terpenes and polyphenols from three species of Oregano (*Lippia graveolens*, *Lippia palmeri* and *Hedeoma patens*) on the suppression of lipopolysaccharide-induced inflammation in RAW 264.7 macrophage cells. *Journal of ethnopharmacology*, *187* (Supplement C), 302-312. doi: https://doi.org/10.1016/j.jep.2016.04.051.

Lima, M. d. S., Quintans-Júnior, L. J., de Santana, W. A., Martins Kaneto, C., Pereira Soares, M. B. & Villarreal, C. F. (2013). Anti-inflammatory effects of carvacrol: Evidence for a key role of interleukin-10. *European Journal of Pharmacology*, *699*(1–3), 112-117. doi: http://doi.org/10.1016/j.ejphar.2012.11.040.

Loizzo, M. R., Menichini, F., Conforti, F., Tundis, R., Bonesi, M., Saab, A. M., . . . Frega, N. G. (2009). Chemical analysis, antioxidant, antiinflammatory and anticholinesterase activities of *Origanum ehrenbergii* Boiss and *Origanum syriacum* L. essential oils. *Food Chemistry*, *117*(1), 174-180. doi: https://doi.org/10.1016/j.foodchem.2009.03.095.

López, P., Sánchez, C., Batlle, R. & Nerín, C. (2007). Vapor-phase activities of cinnamon, thyme, and oregano essential oils and key constituents against foodborne microorganisms. *Journal of Agricultural and Food Chemistry*, *55*(11), 4348-4356. doi: 10.1021/jf063295u.

Ludwiczuk, A., Skalicka-Woźniak, K. & Georgiev, M. I. (2017). Chapter 11 - Terpenoids A2 - Badal, Simone. In R. Delgoda (Ed.), *Pharmacognosy* (pp. 233-266). Boston: Academic Press.

Lukas, B., Schmiderer, C., Mitteregger, U., Franz, C. & Novak, J. (2008). Essential oil compounds of *Origanum vulgare* L.(Lamiaceae) from Corsica. *Molecular and phytochemical analyses of the genus Origanum L.(Lamiaceae)*, *57*(4), 63.

Marrelli, M., Conforti, F., Formisano, C., Rigano, D., Arnold, N. A., Menichini, F. & Senatore, F. (2015). Composition, antibacterial, antioxidant and antiproliferative activities of essential oils from three *Origanum* species growing wild in Lebanon and Greece. *Natural Product Research*, *30*(6), 735-739.

Marrelli, M., Cristaldi, B., Menichini, F. & Conforti, F. (2015). Inhibitory effects of wild dietary plants on lipid peroxidation and on the proliferation of human cancer cells. *Food and Chemical Toxicology*, *86*, 16-24.

Mazzarrino, G., Paparella, A., Chaves-López, C., Faberi, A., Sergi, M., Sigismondi, C., . . . Serio, A. (2015). *Salmonella enterica* and *Listeria monocytogenes* inactivation dynamics after treatment with selected essential oils. *Food Control*, *50* (Supplement C), 794-803. doi: 10.1016/j.foodcont.2014.10.029.

Mechergui, K., Jaouadi, W., Coelho, J. P. & Khouja, M. L. (2016). Effect of harvest year on production, chemical composition and antioxidant activities of essential oil of oregano (*Origanum vulgare* subsp *glandulosum* (Desf.) Ietswaart) growing in North Africa. *Industrial Crops and Products*, *90*, 32-37.

Medzhitov, R. (2008). Origin and physiological roles of inflammation. *Nature*, *454*(7203), 428-435.

Meneses, R., Ocazionez, R. E., Martínez, J. R. & Stashenko, E. E. (2009). Inhibitory effect of essential oils obtained from plants grown in Colombia on yellow fever virus replication in vitro. *Annals of Clinical Microbiology and Antimicrobials*, *8*(1), 8. doi: 10.1186/1476-0711-8-8.

Meyers, C. & Meyers, D. (2008). Thin-layer chromatography. *Current Protocols in Nucleic Acid Chemistry*, A. 3D. 1-A. 3D. 13.

Missopolinou, D., Tsioptsias, C., Lambrou, C. & Panayiotou, C. (2012). Selective extraction of oxygenated compounds from oregano with sub-critical water. *Journal of the Science of Food and Agriculture*, *92*(4), 814-820.

Moraes-Lovison, M., Marostegan, L. F. P., Peres, M. S., Menezes, I. F., Ghiraldi, M., Rodrigues, R. A. F., . . . Pinho, S. C. (2017).

Nanoemulsions encapsulating oregano essential oil: Production, stability, antibacterial activity and incorporation in chicken pâté. *LWT - Food Science and Technology, 77* (Supplement C), 233-240. doi: 10.1016/j.lwt.2016.11.061.

Morshedloo, M. R., Craker, L. E., Salami, A., Nazeri, V., Sang, H. & Maggi, F. (2017). Effect of prolonged water stress on essential oil content, compositions and gene expression patterns of mono- and sesquiterpene synthesis in two oregano (*Origanum vulgare* L.) subspecies. *Plant Physiology and Biochemistry, 111*, 119-128. doi: https://doi.org/10.1016/j.plaphy.2016.11.023.

Ocak, I., Çelik, A., Özel, M. Z., Korcan, E. & Konuk, M. (2012). Antifungal Activity and Chemical Composition of Essential Oil of *Origanum hypericifolium. International Journal of Food Properties, 15*(1), 38-48. doi: 10.1080/10942911003687249.

Ocana-Fuentes, A., Arranz-Gutierrez, E., Senorans, F. & Reglero, G. (2010). Supercritical fluid extraction of oregano (*Origanum vulgare*) essentials oils: anti-inflammatory properties based on cytokine response on THP-1 macrophages. *Food and Chemical Toxicology, 48*(6), 1568-1575.

Öke Altuntaş, F. & Demirtaş, I. (2017). Real-Time Cell Analysis of the Cytotoxicity of *Origanum acutidens* Essential Oil on HT-29 and HeLa Cell Lines. *Turkish Journal of Pharmaceutical Sciences, 14*(1).

Oleszek, W. & Stochmal, A. (2010). High performance liquid chromatography of triterpenes (including saponins). In M. Waksmundzka-Hajnos & J. Sherma (Eds.), *High Performance Liquid Chromatography in Phytochemical Analysis*: CRC Press.

Olmedo, R., Nepote, V. & Grosso, N. R. (2014). Antioxidant activity of fractions from oregano essential oils obtained by molecular distillation. *Food Chemistry, 156*, 212-219. doi: https://doi.org/10.1016/j.foodchem.2014.01.087.

Orhan, I. E., Ozcelik, B., Kartal, M. & Kan, Y. (2012). Antimicrobial and antiviral effects of essential oils from selected Umbelliferae and Labiatae plants and individual essential oil components. *Turkish Journal of Biology, 36*(3), 239-246. doi: 10.3906/biy-0912-30.

Ortega-Nieblas, M. M., Robles-Burgueño, M. R., Acedo-Félix, E., González-León, A., Morales-Trejo, A. & Vázquez-Moreno, L. (2011). Chemical composition and antimicrobial activity of oregano (Lippia palmeri S. WATS) essential oil. *Revista Fitotecnia Mexicana*, *34*(1).

Özbek, H., Güvenalp, Z., Özek, T., Sevindik, H. G., Yuca, H., Yerdelen, K. Ö. & Demirezer, L. Ö. (2017). Chemical composition, antioxidant and anticholinesterase activities of the essential oil of *Origanum rotundifolium* Boiss. from Turkey. *Records of Natural Products*, *11*(5), 9.

Özkan, A. & Erdoğan, A. (2011). A comparative evaluation of antioxidant and anticancer activity of essential oil from *Origanum onites* (Lamiaceae) and its two major phenolic components. *Turkish Journal of Biology*, *35*(6), 735-742.

Ozkan, G., Baydar, H. & Erbas, S. (2010). The influence of harvest time on essential oil composition, phenolic constituents and antioxidant properties of Turkish oregano (*Origanum onites* L.). *Journal of the Science of Food and Agriculture*, *90*(2), 205-209. doi: 10.1002/jsfa.3788.

Ozkan, O. E., Guney, K., Gur, M., Pattabanoglu, E. S., Babat, E. & Khalifa, M. M. (2017). Essential oil of oregano and savory; chemical composition and antimicrobial activity. *Indian Journal of Pharmaceutical Education and Research*, *51*(3), S205-S208. doi: 10.5530/ijper.51.3s.13.

Park, K.-R., Nam, D., Yun, H.-M., Lee, S.-G., Jang, H.-J., Sethi, G., . . . Ahn, K. S. (2011). β-Caryophyllene oxide inhibits growth and induces apoptosis through the suppression of PI3K/AKT/mTOR/S6K1 pathways and ROS-mediated MAPKs activation. *Cancer Letters*, *312*(2), 178-188. doi: http://doi.org/10.1016/j.canlet.2011.08.001.

Pensel, P., Maggiore, M., Gende, L., Eguaras, M., Denegri, M. & Elissondo, M. (2014). Efficacy of essential oils of *Thymus vulgaris* and *Origanum vulgare* on *Echinococcus granulosus*. *Interdisciplinary Perspectives on Infectious Diseases*, *2014*.

Pereira, C. G. & Meireles, M. A. A. (2010). Supercritical fluid extraction of bioactive compounds: fundamentals, applications and economic perspectives. *Food and Bioprocess Technology*, *3*(3), 340-372.

Peschiutta, M., Arena, J., Ramirez Sanchez, A., Gomez Torres, E., Pizzolitto, R., Merlo, C., . . . Zygadlo, J. (2016). Effectiveness of Mexican oregano essential oil from the Dominican Republic (*Lippia graveolens*) against maize pests (*Sitophilus zeamais* and *Fusarium verticillioides*). *Agriscientia*, *33*(2), 89-97.

Pezzani, R., Vitalini, S. & Iriti, M. (2017). Bioactivities of *Origanum vulgare* L.: an update. *Phytochemistry Reviews*, *16*(6), 1253-1268. doi: 10.1007/s11101-017-9535-z.

Piantanida, A. G. & Barron, A. R. (2014). Principles of gas chromatography: OpenStax-CNX.

Pilau, M. R., Alves, S. H., Weiblen, R., Arenhart, S., Cueto, A. P. & Lovato, L. T. (2011). Antiviral activity of the *Lippia graveolens* (Mexican oregano) essential oil and its main compound carvacrol against human and animal viruses. *Brazilian Journal of Microbiology*, *42*, 1616-1624.

Pozzatti, P., Scheid, L. A., Spader, T. B., Atayde, M. L., Santurio, J. M. & Alves, S. H. (2008). In vitro activity of essential oils extracted from plants used as spices against fluconazole-resistant and fluconazole-susceptible Candida spp. *Canadian Journal of Microbiology*, *54*(11), 950-956. doi: 10.1139/W08-097.

Puškárová, A., Bučková, M., Kraková, L., Pangallo, D. & Kozics, K. (2017). The antibacterial and antifungal activity of six essential oils and their cyto/genotoxicity to human HEL 12469 cells. *Scientific Reports*, *7*(1), 8211. doi: 10.1038/s41598-017-08673-9.

Rajan, B., Ravikumar, R., Premkumar, T. & Devaki, T. (2015). Carvacrol attenuates N-nitrosodiethylamine induced liver injury in experimental Wistar rats. *Food Science and Human Wellness*, *4*(2), 66-74. doi: http://doi.org/10.1016/j.fshw.2015.04.002.

Rao, A., Zhang, Y., Muend, S. & Rao, R. (2010). Mechanism of antifungal activity of terpenoid phenols resembles calcium stress and inhibition of the TOR Pathway. *Antimicrobial Agents and Chemotherapy*, *54*(12), 5062-5069. doi: 10.1128/aac.01050-10.

Rassem, H. H., Nour, A. & Yunus, M. (2016). Techniques for extraction of essential oils from plants: a review. *Aus. J. Basic Applied Sci*, *10*(16), 117-127.

Rivero-Cruz, I., Duarte, G., Navarrete, A., Bye, R., Linares, E. & Mata, R. (2011). Chemical composition and antimicrobial and spasmolytic properties of *Poliomintha longiflora* and *Lippia graveolens* essential oils. *Journal of Food Science*, *76*(2), C309-C317. doi: 10.1111/j.1750-3841.2010.02022.x.

Rivero-Cruz, I., Duarte, G., Navarrete, A., Bye, R., Linares, E. & Mata, R. (2011). Chemical composition and antimicrobial and spasmolytic properties of *Poliomintha longiflora* and *Lippia graveolens* essential oils. *Journal of Food Science*, *76*(2).

Rodriguez-Garcia, I., Silva-Espinoza, B. A., Ortega-Ramirez, L. A., Leyva, J. M., Siddiqui, M. W., Cruz-Valenzuela, M. R., . . . Ayala-Zavala, J. F. (2016). Oregano essential oil as an antimicrobial and antioxidant additive in food products. *Critical Reviews in Food Science and Nutrition*, *56*(10), 1717-1727. doi: 10.1080/10408398.2013.800832.

Rodriguez-Garcia, I., Cruz-Valenzuela, M. R., Silva-Espinoza, B. A., Gonzalez-Aguilar, G. A., Moctezuma, E., Gutierrez-Pacheco, M. M., . . . Ayala-Zavala, J. F. (2016). Oregano (*Lippia graveolens*) essential oil added within pectin edible coatings prevents fungal decay and increases the antioxidant capacity of treated tomatoes. *Journal of the Science of Food and Agriculture*, *96*(11), 3772-3778.

Rodríguez-Solana, R., Salgado, J. M., Domínguez, J. M. & Cortés-Diéguez, S. (2015). Comparison of soxhlet, accelerated solvent and supercritical fluid extraction techniques for volatile (GC–MS and GC/FID) and phenolic compounds (HPLC–ESI/MS/MS) from Lamiaceae Species. *Phytochemical Analysis*, *26*(1), 61-71.

Rubió, L., Motilva, M.-J. & Romero, M.-P. (2013). Recent advances in biologically active compounds in herbs and spices: a review of the most effective antioxidant and anti-inflammatory active principles.

Critical Reviews in Food Science and Nutrition, 53(9), 943-953. doi: 10.1080/10408398.2011.574802.

Sánchez, C., Aznar, R. & Sánchez, G. (2015). The effect of carvacrol on enteric viruses. *International Journal of Food Microbiology, 192* (Supplement C), 72-76. doi: 10.1016/j.ijfoodmicro.2014.09.028.

Santoyo, S., Jaime, L., García-Risco, M. R., Ruiz-Rodríguez, A. & Reglero, G. (2014). Antiviral properties of supercritical CO2 extracts from oregano and sage. *International Journal of Food Properties, 17*(5), 1150-1161. doi: 10.1080/10942912.2012.700539.

Scandorieiro, S., de Camargo, L. C., Lancheros, C. A. C., Yamada-Ogatta, S. F., Nakamura, C. V., de Oliveira, A. G., . . . Kobayashi, R. K. T. (2016). Synergistic and additive effect of oregano essential oil and biological silver nanoparticles against multidrug-resistant bacterial strains. *Frontiers in Microbiology, 7*(760). doi: 10.3389/fmicb.2016.00760.

Schmidt, E., Bail, S., Buchbauer, G., Stoilova, I., Krastanov, A., Stoyanova, A. & Jirovetz, L. (2008). Chemical composition, olfactory evaluation and antioxidant effects of the essential oil of *Origanum majorana* L. from Albania. *Natural Product Communications, 3*(7), 1051-1056.

Sellami, I. H., Maamouri, E., Chahed, T., Wannes, W. A., Kchouk, M. E. & Marzouk, B. (2009). Effect of growth stage on the content and composition of the essential oil and phenolic fraction of sweet marjoram (*Origanum majorana* L.). *Industrial Crops and Products, 30*(3), 395-402. doi: 10.1016/j.indcrop.2009.07.010.

Silva, F. V., Guimarães, A. G., Silva, E. R. S., Sousa-Neto, B. P., Machado, F. D. F., Quintans-Júnior, L. J., . . . Oliveira, R. C. M. (2012). Anti-inflammatory and anti-ulcer activities of carvacrol, a monoterpene present in the essential oil of oregano. *Journal of Medicinal Food, 15*(11), 984-991. doi: 10.1089/jmf.2012.0102.

Soares, I. H., Loreto, É. S., Rossato, L., Mario, D. N., Venturini, T. P., Baldissera, F., . . . Alves, S. H. (2015). In vitro activity of essential oils extracted from condiments against fluconazole-resistant and -sensitive

Candida glabrata. *Journal de Mycologie Médicale*, *25*(3), 213-217. doi: 10.1016/j.mycmed.2015.06.003.

Soto-Armenta, L., Sacramento-Rivero, J., Acereto-Escoffié, P., Peraza-González, E., Reyes-Sosa, C. & Rocha-Uribe, J. (2017). Extraction yield of essential oil from *Lippia graveolens* leaves by steam distillation at laboratory and pilot scales. *Journal of Essential Oil Bearing Plants*, *20*(3), 610-621.

Sowa, I., Wójciak-Kosior, M., Rokicka, K., Kocjan, R. & Szymczak, G. (2014). Application of solid phase extraction with the use of silica modified with polyaniline film for pretreatment of samples from plant material before HPLC determination of triterpenic acids. *Talanta*, *122* (Supplement C), 51-57. doi: https://doi.org/10.1016/ j.talanta.2014.01.039.

Spagnoletti, A., Guerrini, A., Tacchinia, M., Vinciguerra, V., Leone, C., Maresca, I., . . . Angionella, L. (2016). Chemical composition and bio-efficacy of essential oils from italian aromatic plants: *Mentha suaveolens, Coridothymus capitatus, Origanum hirtum* and *Rosmarinus officinalis*. *Natural Product Communications*, *11*, 1517-1520.

Téllez, L., Arévalo, F., Juárez, H., Altamirano, P., Ccapa, K., Chávez, J. & Visitación, L. (2014). Determinación de timol y carvacrol en hojas de orégano por HPLC FL. *Revista de la Sociedad Química del Perú*, *80*(4), 279-286. [Thymol and carvacrol HPLC determination from oregano leaves. *Journal of the Peruvian Chemistry Society*. *80*(4), 279-286].

Tepe, B., Cakir, A. & Sihoglu Tepe, A. (2016). Medicinal uses, phytochemistry, and pharmacology of *Origanum onites* (L.): A Review. *Chemistry & Biodiversity*, *13*(5), 504-520. doi: 10.1002/cbdv.201500069.

Tongnuanchan, P. & Benjakul, S. (2014). Essential oils: extraction, bioactivities, and their uses for food preservation. *Journal of Food Science*, *79*(7).

Trinchieri, G. (2012). Cancer and inflammation: an old intuition with rapidly evolving new concepts. *Annual Review of Immunology, 30*, 677-706. doi: 10.1146/annurev-immunol-020711-075008.

Tuttolomondo, T., La Bella, S., Licata, M., Virga, G., Leto, C., Saija, A., . . . Napoli, E. M. (2013). Biomolecular characterization of wild Sicilian oregano: phytochemical screening of essential oils and extracts, and evaluation of their antioxidant activities. *Chemistry & Biodiversity, 10*(3), 411-433.

Ultee, A., Bennik, M. H. J. & Moezelaar, R. (2002). The phenolic hydroxyl group of carvacrol is essential for action against the food-borne pathogen *Bacillus cereus*. *Applied and Environmental Microbiology, 68*(4), 1561-1568. doi: 10.1128/aem.68.4.1561-1568.2002.

Vale-Silva, L., Silva, M.-J., Oliveira, D., Gonçalves, M.-J., Cavaleiro, C., Salgueiro, L. & Pinto, E. (2012). Correlation of the chemical composition of essential oils from *Origanum vulgare* subsp. virens with their in vitro activity against pathogenic yeasts and filamentous fungi. *Journal of Medical Microbiology, 61*(2), 252-260. doi: 10.1099/jmm.0.036988-0.

Van Haute, S., Raes, K., Van der Meeren, P. & Sampers, I. (2016). The effect of cinnamon, oregano and thyme essential oils in marinade on the microbial shelf life of fish and meat products. *Food Control, 68* (Supplement C), 30-39. doi: 10.1016/j.foodcont.2016.03.025.

Vazirian, M., Mohammadi, M., Farzaei, M., Amin, G. & Amanzadeh, Y. (2015). Chemical composition and antioxidant activity of *Origanum vulgare* subsp. *vulgare* essential oil from Iran. *Research Journal of Pharmacognosy, 2*(1), 41-46.

Velasco, R. J., Villada, H. S. & Carrera, J. E. (2007). Aplicaciones de los fluidos supercríticos en la agroindustria. *Información tecnológica, 18*(1), 53-66. [Agroindustry applications of the supercritical fluids. *Technological Information*. 18(1), 53-66).

WHO. (2017). Cancer Fact Sheet. Retrieved November 2017, from http://www.who.int/mediacentre/factsheets/fs297/en/.

Witte, L. (1986). GC-MS methods for terpenoids. *Gas Chromatography/Mass Spectrometry* (pp. 134-145): Springer.

Yadav, N., Yadav, R. & Goyal, A. (2014). Chemistry of terpenoids. *International Journal of Pharmaceutical Sciences Review and Research*, *27*(2), 272-278.

Yaldiz, G., Arici, Y. K. & Yilmaz, G. (2017). Phytochemical analysis, antioxidant and antibacterial activities of four *Lamiaceae* species Cultivated in Barnyard Manure. *Tarım Bilimleri Dergisi*, *23*(1), 95-108.

Yan, F., Azizi, A., Janke, S., Schwarz, M., Zeller, S. & Honermeier, B. (2016). Antioxidant capacity variation in the oregano (*Origanum vulgare* L.) collection of the German National Genebank. *Industrial Crops and Products*, *92*, 19-25.

Zabot, G. L., Moraes, M. N., Rostagno, M. A. & Meireles, M. A. A. (2014). Fast analysis of phenolic terpenes by high-performance liquid chromatography using a fused-core column. *Analytical Methods*, *6*(18), 7457-7468.

Zheljazkov, V. D., Astatkie, T. & Schlegel, V. (2012). Distillation time changes oregano essential oil yields and composition but not the antioxidant or antimicrobial activities. *HortScience*, *47*(6), 777-784.

In: Terpenes
Editor: Albin Bjarke
ISBN: 978-1-53613-622-7

Chapter 2

Conversion of Monoterpenes into Flavors and Fragrances over Solid Acid Catalysts

J. E. Castanheiro *

Centro de Química de Évora, Departamento de Química,
Universidade de Évora, Évora, Portugal

Abstract

Monoterpenes are renewable resources that are widely used in the pharmaceutical, cosmetic and food industries as active components of drugs and ingredients of artificial flavours and fragrances. Acids catalyzed of terpenes are important synthetic routes to valuable compounds. Traditionally, strong homogeneous acid catalysts, e.g., sulphuric acid, have been used in acid reactions. However, solid acids have many advantages over liquid acids. They do less harm to the environment and do not have corrosion or effluent disposal problems. They are reusable and easy to be separated from liquid products.

This review focuses on solid acid catalysts as potential heterogeneous catalysts for terpenes conversion into fragrances and flavours, replacing homogenous ones. Different reactions, such as

* Corresponding Author Email: jefc@uevora.pt.

hydration, alkoxylation, esterification, isomerization and cyclization of terpenes over heterogenous catalyts will be reviewed.

Keywords: terpenes, heterogenous catalysts, solid acid, flavours and fragrances

INTRODUCTION

Monoterpenoids are important, renewable raw materials used for the synthesis of aroma chemicals. Mixed terpene feedstock is a valuable by product from cellulose derived materials and has a potential for industrial organic synthesis [1-4]. Catalytic terpene transformations have been carried out over heterogeneous catalysts. The isomerization reaction of alpha-pinene, usually performed in the liquid phase, has been studied over several different heterogenous catalysts [2], yielding limonene and camphene as the main products.

Other compounds, like borneol, isoborneol and their acetates, usually obtained from camphene, are used in formulations of perfumes and medicines, as well as in the industrial production of synthetic camphor. These reactions have been studied over solid acid catalysts.

Alpha-terpineol and alpha-terpinyl acetate, which is the (commercially) most interesting monocyclic terpenic alcohol, can be obtained by alpha-pinene, beta-pinene and limonene hydration and acetoxylation, respectively [1, 3].

The alkoxylation of alpha-pinene, beta-pinene and limonene yield a mixture of terpenic ethers which exhibit organoleptic properties, and thus, they are used in various fragrance compositions. Also, terpenic alcohols, aldehydes, and esters exhibit organoleptic properties [1, 4].

In this work, will be made a review of terpenes transformations, such as, alkoxylation, hydration, esterification, isomerization and cyclization, over solid acid catalysts.

CONVERSION OF TERPENES

Alkoxylation of Terpenes

The alkoxylation of alpha-pinene and limonene with Cl-C5 alcohols to 1-methyl-4-[alpha-alkoxy-isopropyl]-l-cyclohexene was carried out over beta zeolite in both a discontinuous batch reactor and a continuous flow-type apparatus with a fixed bed reactor [5]. In both reactors, the use of methanol as an addition compound and limonene as feedstock gives 1-methyl-4-[alpha-methoxy-isopropyl]-l-cyclohexene with the yield of 85% (conversion: 93%, selectivity: 92%). By means of variation of the reaction parameters, the limonene conversion can be adjusted within the range 40-90%. The selectivity to 1-methyl-4-[alpha-methoxy-isopropyl]-l-cyclohexene remains always about 95%. For mixtures of methanol and alpha-pinene, the highest yield of corresponding alpha-terpinyl methyl ether is only of about 50% (conversion: 99%, selectivity: 51%) in both reactor types, as several bicyclic and double addition products are formed in parallel. The selectivity of all addition products can reach values of about 85%. In the case of limonene and alpha-pinene alkoxylation with higher alcohols (C2-C5 alcohols), good results are obtained by using linear alcohols only. Other (iso- and branched) alcohols tend to dehydrate to corresponding alkenes and water. This might also reflect a shape selective effect. Alkoxylation of alpha-pinene or limonene in the presence of other large pore zeolites, e.g., H-USY or H-mordenite, and solid acids, e.g., ion exchanger or heteropolyacids, results in a considerable selectivity reduction to the corresponding alpha-terpinyl alkyl ethers. The destruction of the structure of beta zeolite leads to a complete deactivation of the catalyst. The beta zeolite is more active as well as more selective than all other catalysts described in the literature for this reaction type. The most likely reasons for the high activity of beta zeolite are the relatively high BET surface area of the catalyst and the large amount of acid sites measured by temperature-programmed desorption (TPD) of ammonia [5].

Methoxylation of alpha-pinene was studied over sulfonic acid-functionalized mesoporous silica (MCM-41, PMO) at 60ºC [6]. The support functionalization was achieved by the introduction of 3-(mercaptopropyl)trimethoxysilane onto the surface of these materials either by grafting or by co-condensation. The thiol groups were oxidized to SO_3H by treatment with H_2O_2. All the catalysts were active in the studied reaction, PMO-SO3H-g being the best one. Good values of selectivity to alpha-terpinyl methyl ether were obtained with these catalysts. Catalytic stability of the PMO-SO_3H-g was evaluated by performing consecutive batch runs with the same catalyst sample. After the third batch, a stabilization of the activity was observed [6].

Methoxylation of alpha-pinene was studied using poly(vinyl alcohol) (PVA) containing sulfonic acid groups as catalysts [7]. These groups were introduced by the direct reaction between –OH groups of PVA and sulfosuccinic acid (SSA). Poly(vinyl alcohol) catalysts were prepared using sulfosuccinic acid (SSA) by varying the amount of SSA (5-40 mol%). It was observed that the amount of acid sites increases with the amount of SSA. It was also observed that the diffusivity of alpha-pinene through the polymeric catalysts decreases with the increase of the crosslinking degree. The conversion of alpha-pinene increases when the amount of sulfosuccinic acid used in the polymer crosslinking is increased from 5% to 40%. However, when the crosslinking degree increases from 20% to 40%, the conversion of alpha-pinene increases only slightly. Good values of selectivity to alpha-terpinyl methyl ether (about 60% near complete conversion) were obtained over poly(vinyl alcohol) with sulfonic acid groups. Catalytic stability of the PVA_SSA20 was evaluated by performing consecutive batch runs with the same catalyst sample. After the third batch, a stabilization of the initial activity was observed. A kinetic model was developed assuming that the alpha-pinene is consumed according to the parallel reaction network. It was observed that the kinetic model fits the experimental concentration data quite well [7].

Dodecatungstophosphoric acid, dodecamolybdophosphoric, and tungstosilicic acid and molybdosilicic acid immobilized on silica were used in the alpha-pinene methoxylation [8]. The main product of alpha-pinene methoxylation was alpha-terpinyl methyl ether. Also being formed was bornyl methyl ether, fenchyl methyl ether, limonene and terpinolene as by-products. It was observed that the catalytic activity increases in the series: SiMo < PMo_S < SiW_S < PW2_S. Catalysts consisting of the PW immobilized on silica with different PW loading were prepared. It was observed that the catalytic activity increases with the amount of PW immobilized on silica. However, at a high amount of PW on silica, a decrease of the catalytic activity was observed. All the catalysts tested on alpha-pinene methoxylation exhibited good selectivity (about 60% at near complete alpha-pinene conversion). To optimize the reaction conditions, the effect of various parameters, such as reaction temperature, molar ratio of a-pinene to methanol and catalyst loading of PW2_S were carried out. The catalytic stability of the PW2_S was evaluated by performing consecutives batch runs with the same catalyst sample. Similar activity for all the catalytic tests was observed. A simple kinetic model, which fits the experimental concentration data quite well, has been developed [8].

The acid-catalyzed alkoxylation of alpha-pinene, beta-pinene and limonene with Cl-C4 alcohols (methanol, ethanol, 1-propanol and 1-buthanol) to alpha-terpinyl alkyl ether was carried out in the presence of silica-occluded heteropolyacid (tungstophosphoric acid) [9]. This ether is used as flavors and fragrances for perfume and cosmetic products, as additives for pharmaceuticals and agricultural chemicals, as well as in the food industry for organic synthesis. For the alkoxylation of alpha-pinene, beta-pinene and limonene, the catalytic activity decreases with an increase in the chain length of the linear alcohols. This behavior could be explained due to the presence of sterical hinderance and diffusion restrictions in the porous system of the catalyst. The selectivity of silica-occluded heteropolyacid to alpha-terpinyl alkyl ether also decreased with the increase of the chain length of the alcohols. The catalytic stability of the PW-S was studied in the ethoxylation of alpha-pinene and beta-pinene. It was observed that the catalyst can easily be recovered and reused without loss of activity. It can be also observed that the alkoxylation reactions with beta-pinene substrate were faster than the alpha-pinene and limonene [9].

A biomass derived carbon, a commercial microporous carbon and a xerogel mesoporous carbon catalyst were applied in the study of alpha-pinene methoxylation reaction [10]. The results revealed that from the biomass waste used (olive stones), it is possible to obtain a microporous carbon which can later be used as an acid catalyst, resulting in a clear added value to the material. The catalytic activity increased in the order Cnorit< CMN< CB. The commercial catalyst was found to be the less suitable one. The results also seem to indicate that the most important feature of these catalysts is the surface chemistry, although some effect of the textural properties could be observed. All catalysts showed good selectivity towards the alpha-terpinyl methyl ether. These results show that it is possible to reutilize a waste residue and convert it to high added value product such as an acid catalyst [10].

Alkoxylation of alpha-pinene, beta-pinene and limonene was performed in the presence of SBA-15-occluded tungstophosphoric acid (HPW) [11]. The HPW was immobilized in SBA-15 using the sol-gel method. The catalysts were characterized by N_2 adsorption, FT-IR, Raman spectroscopy, X-Ray diffraction, ICP-AES and TEM. A series of catalysts with different heteropolyacid loadings ranging from 1.8 to 19.3 wt.% were prepared. PW4-SBA-15 (with 10.8 wt.%) exhibited the highest catalytic activity for the alkoxylation of alpha-pinene with ethanol. An approximately 53% selectivity to alpha-terpinyl ethyl ether was observed over the PW-SBA-15 catalysts. PW4-SBA-15 was also used as a catalyst for the alkoxylation of other terpenes, including -pinene and limonene. The catalytic activity decreased in the following order: beta-pinene > alpha-pinene > limonene. The PW4-SBA-15 catalyst exhibited high catalytic stability for the alkoxylation of alpha-pinene [11].

The alkoxylation of camphene with C1-C4 alcohols (methanol, ethanol, 1-propanol, 2-propanol, 1-butanol, and 2-butanol) to alkyl isobornyl ether was successfully carried out in the presence of silica-occluded $H_3PW_{12}O_{40}$ [12]. The catalytic activity decreases with an increase of the chain length of the linear alcohols, which can be explained due to the presence of diffusion restrictions in the porous system of catalyst and sterical hindrance. High selectivity for alkyl isobornyl ether was obtained in the presence of silica-occluded heteropolyacid. The catalyst (PW-Ssg) can be recovered and reused without loss of activity and selectivity. The stability of PW-Ssg catalyst towards metal leaching (less than 1%) is much greater than that of PW-Sim which lost 20% of its initial amount of heteropolyacid, during the catalytic reaction [12].

The alkoxylation of camphene with methanol, ethanol and 1-butanol was studied in the presence of $H_3PW_{12}O_{40}$ and $Cs_{2.5}H_{0.5}PW_{12}O_{40}$ [13]. In the same work, de Meireles et al. [13] studied the alkoxylation of alpha-pinene, beta-pinene and limonene with 1-butanol over $H_3PW_{12}O_{40}$ and $Cs_{2.5}H_{0.5}PW_{12}O_{40}$. The authors observed that the heteropolyacid $H_3PW_{12}O_{40}$ and its acidic Cs salt, $H_3PW_{12}O_{40}$, are efficient catalysts for the liquid phase alkoxylation of camphene, limonene, alpha-pinene, and beta-pinene. These reactions provide effective routes for the synthesis of isobornyl and alpha-terpenyl ethers [13].

Acetoxylation and Hydration

The liquid-phase hydration and acetoxylation of the dihydromyrcene are carried out in the presence of $H_3PW_{12}O_{40}$ (HPA) at 14–30°C to yield dihydromyrcenol and dihydromyrcenyl acetate [14]. Homogeneous, biphasic and heterogeneous (with HPA/SiO_2 catalyst) reactions have been studied. HPA has a much higher catalytic activity than conventional acid catalysts such as H_2SO_4 and Amberlyst-15. A synthetically useful biphasic acetoxylation/hydration of DHM has been developed. The method provides a 90% selectivity to the sum of dihydromyrcenol and dihydromyrcenyl acetate at 21% DHM conversion and allows for easy and clean catalyst recycling without loss of activity [14].

The hydration of dihydromyrcene (DHM) to produce dihydromyrcenol (DHM-OH) was carricd out in the presence of H-beta zeolite, as heterogenous catalyst [15]. By using a non-protic co-solvent like acetone to work under a single-phase system, a reasonable conversion with excellent selectivity to DHM-OH is obtained, although some double bond isomerization of the terpene is also taking place. Non-protic co-solvents work better than protic ones, presumably since the last ones compete with water for adsorption on Brönsted acid sites. The reaction is not controlled by intrapore diffusion, and samples with different crystal sizes gave similar activities and selectivities. However, defect free beta samples with high hydrophobic character are preferred for the reaction due to their improved adsorption properties for non-polar molecules. Working with those materials, a more than 50% yield of DHM-OH was obtained with a selectivity close to 100%. The influence of zeolite hydrophobicity was further proved in a two-phase liquid reaction system. In such conditions, best results were achieved with the most hydrophobic catalysts [15].

Acetoxylation of alpha-pinene was carried out in the presence of H-beta zeolites at 20°C [16]. The main product of alpha-pinene acetoxylation was alpha-terpinyl acetate. The rate of reaction of the alpha-pinene is directly related to the zeolite SiO_2/Al_2O_3 ratio. Dealumination of the catalysts by acid treatment slightly improves reaction time. Increasing the temperature is detrimental, leading to an increase in the bornyl acetate concentration. The zeolites can be recycled [16].

Hydration and acetoxylation of limonene and alpha-pinene into terpineol and terpinyl acetate was carried out in the presence of transition metal and rare earth ion-exchanged beta zeolite [17]. These catalysts, under optimized reaction conditions, showed higher activity and selectivity compared to conventionally used acid catalysts such as H_2SO_4 and amberlyst-15. Conversion levels were 9–26% and 58–82% for limonene in the presence of glacial and aqueous acetic acid, respectively, and the selectivity for major products alpha-terpinyl acetate and terpineol were up to 54% and 65%, respectively. Conversion values in the range of 62–100% and 72–100% were obtained for α-pinene in the presence of glacial and aqueous acetic acid, respectively. From the measured acidity data of these zeolites, it is observed that both hydration and acetoxylation are Brönsted acid-catalyzed reactions. Isomerisation and addition reactions were found simultaneously in the system and bornyl acetate and alpha-terpineol were the major products among addition products for glacial and aqueous acetic acid, respectively. In the conversion of both limonene and α-pinene the aqueous acetic acid conditions are more favorable. The study shows the potential of ion-exchange beta zeolites for Brönsted acid-catalyzed monoterpene transformation [17].

The acetoxylation of alpha-pinene was carried out over SBA-15 with sulfonic acid groups [18]. The products of acetoxylation of alpha-pinene are acetates (alpha-terpinyl acetate, bornyl acetate and alpha-fenchyl acetate) and hydrocarbons (camphene, tricyclene, limonene, gamma-terpinene, terpinolene and alpha-terpinene). Catalysts with different amounts of sulfonic acid groups were prepared. The conversion increased with the increase of surface area and porous volume of the catalysts, probably due to the increases of alpha-pinene accessibility to the active sites. The introduction of the sulfonic groups promotes the conversion to the desired products. To optimize the reaction conditions, different parameters, such as catalyst amount, initial concentration of alpha-pinene and temperature were studied. The catalytic stability of C1 sample was also studied. After the second use, the catalytic activity of C1 remained constant [18].

The liquid-phase hydration and acetoxylation of limonene, alpha-pinene, and beta-pinene catalyzed by dissolved or silica-supported heteropoly acid $H_3PW_{12}O_{40}$ (PW) in acetic acid and acetic acid/water solutions have been studied [19]. All three substrates give alpha-terpineol as the main product along with alpha-terpenyl acetate. The reaction rate increases in the order: limonene $<$ alpha-pinene $<$ beta-pinene. Synthetically useful homogeneous and heterogeneous acetoxylation and hydration of limonene, alpha-pinene and beta-pinene into alpha-terpineol and alpha-terpenyl acetate have been developed. At room temperature under optimized conditions, alpha-pinene and beta-pinene form a mixture of alpha-terpineol and alpha-terpenyl acetate with 85% selectivity at 90% substrate conversion. Limonene gives alpha-terpineol and alpha-terpenyl acetate with 85% selectivity at 50% conversion, with the main product being ester 5 in acetic acid (4/5 $\approx$ 30/70) and alcohol 4 in HOAc/H2O (90/10 v/v) solutions (4/5 $\approx$ 85/15). No oligomerization of monoterpenes occurs under the optimized conditions. The catalyst can be separated without neutralization and may be reused. The PW shows a much higher catalytic activity than conventional acid catalysts such as H_2SO_4 and Amberlyst-15 [19].

The liquid-phase acetoxylation and acetoxylation/hydration of camphene catalyzed by dissolved or silica-supported heteropoly acid $H_3PW_{12}O_{40}$ (PW) in AcOH/H2O solutions give isobornyl acetate and isoborneol with an almost 100% total selectivity. PW shows a much higher catalytic activity than conventional acid catalysts such as H2SO4 and Amberlyst-15. PW can be recovered without neutralization and reused [20].

The esterification of camphene with carboxylic acids (C2, C4 and C6 short-chain fatty acids) was performed using $H_3PW_{12}O_{40}$ (PW) supported on silica. This reaction is an effective route for the synthesis isobornyl carboxylates, potential fragrances possessing a strong characteristic aroma, starting from a renewable biomass-based substrate. The reaction yield is equilibrium-controlled, with an ester selectivity of virtually 100%. Under optimized conditions, the esters have been obtained in 80–90% yield and up to 3000 PW turnover number. The catalyst can be recovered and reused without loss of activity and selectivity [21].

The acetoxylation of camphene was carried out in the presence of Beta zeolite with Si/Al ratio equal to 40 [22]. The main product of camphene acetoxylation was isobornyl acetate, also forming bornyl acetate and fenchyl acetate. The effects of various parameters, such as reactant molar ratio, reaction temperature, catalysts loading, and reusability of the catalyst were studied to optimize the reaction conditions. The catalytic stability of beta zeolite was studied by performing consecutive batch runs with the same catalyst sample. Similar values of the catalytic activity were observed after the third batch runs [22].

The hydration of α-pinene and camphene was carried out in the presence of USY catalysts in aqueous acetone at 55°C [23]. The main product of α-pinene hydration is α-terpineol while the main product of camphene hydration is isoborenol. The catalyst samples were prepared from zeolite Y by hydrothermal treatments at temperatures ranging from 450 to 850°C. The extra-framework aluminium species were kept in the samples. Although zeolite H-Y is not active as a catalyst for these hydration reactions, USY catalysts show a reasonable activity and are very selective to the above mentioned terpenic alcohols. Selectivity increases with the relative concentration of Lewis sites [23].

The acid catalyzed hydration of alpha-pinene yields a complex mixture of monoterpenes (alcohols and hydrocarbons) [24]. By controlling the reaction variables it is possible to make it selective to the terpenic alcohols, namely α-terpineol. In this chapter the results of the hydration reaction of alpha-pinene catalysed by polydimethylsiloxane (PDMS) membranes filled with zeolite USY, zeolite beta or a surface modified activated carbon, are reported. The reaction is carried out at 50°C, using aqueous acetone as solvent. The activity and selectivity towards α-terpineol of the composite membranes are compared with those of the free catalysts. For all the catalysts tested, the composite membranes exhibit lower activities than the free catalysts. When the catalyst loading is increased, the membrane activity increases, probably due to the increase of the membrane channel system. The PDMS membrane seems to have complex effects on the selectivity to alpha-terpineol. It increases for the USY zeolite, decreases for the beta and shows only a slight increase for the activated carbon. The effect of the solvent water content on the selectivity as well as the effect of the particle size on the membrane sorption properties may explain those different behaviors [24].

The main reaction product of the hydration of alpha-pinene in the presence of a catalytic PDMS membrane loaded with a USY zeolite is alpha-terpineol, simultaneously forming several minor products, mainly terpenic hydrocarbons [25]. The concentration profiles of reagents and products exhibit a pronounced initial induction period followed by a rapid increase of the reaction rate. This feature suggests an autocatalytic behavior. Alpha-pinene permeation measurements across the membrane show that alpha-pinene molar flux depends almost linearly of pinene concentration in the range of relatively low concentrations. However, it increases strongly in the presence of increasing amounts of alpha-terpineol. A simple diffusion-kinetic model, which fits experimental concentration data quite well, has being developed. It assumes a parallel reaction network and an exponential dependence of pinene diffusivity in the membrane on alpha-terpineol concentration. The value of the alpha-pinene initial diffusivity obtained from the model increases with the catalyst loading. Simultaneously, the ratio of the model's apparent kinetic constants, k1/k2, increases for the highest catalyst loading, reflecting an increase in selectivity to alpha-terpineol. This behavior shows that the increase in the catalyst loading leads to an increase in the permeability not only of alpha-pinene but also of water, probably due to an increase in the membrane channeling [25].

Polymeric catalytic membranes consisting of HPMo entrapped in PVA cross-linked with succinic acid are active for the alpha-pinene hydration reaction [26]. Membrane activity increases with the polymer cross-linking due to the increase in membrane hydrophobicity. However, this growth in activity is limited by the increasing restrictions to the molecules mobility. It was not possible to prepare polymeric catalytic membranes consisting of HPMo directly entrapped in a PDMS matrix, probably due to interference of HPA with the polymer cross-linking. Alternatively, the preparation of a composite membrane consisting of USY encaged HPMo dispersed in a PDMS matrix was successful since HPA is mostly retained inside the zeolite supercages. This membrane shows a surprisingly higher activity for pinene hydration, which is even higher than that observed with USY encaged HPMo alone. However, selectivity to alpha-terpineol is slightly lower than that obtained for the HPMo/PVA cross-linked with 20 or 30% of succinic acid. Membrane stability seems to be reasonably good. For the HPMo/PVA membranes, an initial loss in HPA is observed, but in the subsequent runs, the amount of HPA present in the membrane seems to be stabilized. For the HPMo-USY/PDMS membrane, the observed losses in HPA are much lower than those observed for the HPMo/PVA membranes, showing that the HPA zeolite encapsulation is effective in the catalyst stabilization. In both cases, HPMo/PVA or HPMo-USY/PDMS, catalyst activity increases with the subsequent uses of the same membrane, probably due to the interaction between retained alpha-terpineol and the polymer matrix, promoting a better diffusion of pinene molecules to the catalyst active sites [26].

The hydration reaction of alpha-pinene was carried out in the presence of a poly(vinyl alcohol)/molybdophosphoric acid (PVA/HPMo) membrane crosslinked with succinic acid and modified by treatment with acetic anhydride in order to improve its hydrophobic properties. The increase of membrane hydrophobicity with acetylation is documented not only by the water droplet contact angle but also by the sorption coefficients of α-pinene and water. The introduction of acetyl groups improves the membrane transport properties, as reflected by pinene diffusivity calculated from permeation data. A kinetic-diffusion model was developed assuming that the reaction product alpha-terpineol affects the transport of water and α-pinene across the acetylated PVA membranes. When membrane acetylation increases, the model predicted kinetic constants for hydration and isomerization reactions decrease, although the initial water diffusivity increases. These results suggest that the increase of catalytic activity may be due to an improvement of water transport across the membrane [27].

The hydration reaction of alpha-pinene in the presence of natural clays treated with monochloroacetic acid as catalyst to obtain oxygenated compounds was studied [28]. Catalysts were characterized by different techniques, such as XRD, DTA, TPD of adsorbed pyridine, and FTIR of adsorbed pyridine to determine Brønsted and Lewis acid sites. The natural clay, without treatment, produced compounds resulting from α-pinene isomerization, whereas the treated clays produced alcohols and other products in addition to isomerization compounds. The acid treatment increased the acid strength of the natural clay and resulted in Brønsted acid centres. Catalytic tests showed that treatment of the natural clay with the acid improved the catalytic activity and the selectivity toward oxygenated products by increasing the acidity of the catalyst. Alpha-pinene conversions of approximately 80% with selectivities toward oxygenated compounds of 70% were observed, where alpha-terpineol was the major component. The selectivity of oxygenated products increased with the degree of conversion because of the greater availability of water molecules to the acid sites. Catalytic stability was studied, and the percent conversion decreased when the number of reuses increased. However, the selectivity toward oxygenated compounds increased [28].

The hydration of turpentine over Amberlyst-15 wet was successfully carried out in the stirred tank reactor [29]. During the hydration process, the isomerization of alpha-pinene also took place under the catalytic condition. The effects of different parameters such as agitation speed, catalyst type, catalyst loading, mass ratio of the reactants, temperature and the reusability of catalyst on the conversion of turpentine and the yield of alpha-terpineol were investigated to obtain the optimal reaction conditions in a stirred tank reactor. Based on these optimized conditions, the experiment was conducted in a new pilot-scale jet reactor to achieve an excellent performance. In this system, the four reactions were parallel. A pseudo-homogeneous (PH) model was used to explain the reaction rate equation and to determine the reaction rate constants and reaction activation energy. Kinetic study of these reactions indicates that the reaction rate is slow in the temperature range 328.15–348.15 K. The activation energy of the four reactions was determined to be 66.05 kJ/mol, 95.05 kJ/mol, 83.07 kJ/mol and 85.71 kJ/mol, respectively. Under the optimal conditions, the experiment was conducted again using a new pilot-scale jet reactor to achieve excellent performance [29].

The hydration of alpha-pinene using solid acid catalysts was studied [30]. The catalysts were prepared by impregnating trichloroacetic acid (TCA) on different supports such as silica, titania and zirconia (TCA/SiO_2, TCA/TiO_2 and $TCA/ZrO_2.nH_2O$, respectively). The materials were characterized by isotherms of nitrogen at 77 K, FT-IR, TGA, XPS, XRD. The conversion of α-pinene with $TCA/ZrO_2.nH_2O$ was 5.7-times higher than that observed for TCA/TiO_2. The adsorption of α-pinene over TCA/ZrO_2nH_2O was 5.2-times higher than over TCA/TiO_2. The TCA/TiO_2 catalyst converted alpha-pinene into hydrocarbons, while the $TCA/ZrO_2.nH_2O$ catalyst was active and selective for producing alcohols, with a conversion of 57% and a selectivity of 75% of total alcohols, and showed 57% selectivity for alpha-terpineol. The TCA/SiO_2 sample did not show catalytic activity due to the elimination of the trichloroacetic acid during the preparation step of the catalyst. In the case of $TCA/ZrO_2.nH_2O$, used and reused catalysts exhibited lower catalytic activities, but higher selectivities for alcohols. The decrease in catalytic activity is related to a decrease in the Cl/Zr mass ratio on the surface of the catalyst [30].

Isomerization

The isomerization of alpha-pinene to camphene and limonene was carried out over acid clays (dioctahedral and trioctahedral) [31]. The specific surface area and the pore volume increase and the total acidity in the clay by the acid treatment enhances the performance for its use in the conversion of alpha-pinene. Conversions of alpha-pinene of about 50% with selectivities in camphene between 52.6–54.5% and selectivities in limonene between 37.6–44.6%, were obtained with acid-treated dioctahedral and trioctahedral smectites after 1h of reaction at 373K. The catalytic activity of smectites for the alpha-pinene conversion was correlated with the structural and textural characteristics, and the total acidity of the clays [31].

The isomerization of alpha-pinene was studied over different pre-treated montmorillonite clays [32]. The effects of the acid pre-treatment parameters (acid-to-clay ratio (A/C), water-to-acid ratio (W/A), temperature and pre-treatment time) on the product distribution were investigated. It was observed that the pre-treatment with acid improved the catalytic activity of that catalyst which was activated at 200° C for 4h at A/C ratios between 0.2 and 0.4. Alpha-pinene consumption was very small over the catalysts prepared with an A/C ratio of 0.8. High A/C ratio such as 0.8 causes drastic changes in the structure and considerable drops in the activity of the catalysts are observed. The A/C and W/C ratios directly affect the dealumination degrees of catalysts corresponding to changes in the extent of acidity of the catalysts. The differences created in the distribution of Brönsted and Lewis acid sites in catalysts by acid treatment cause significant changes on product distributions in alpha-pinene isomerization. Catalysts with acid sites, of which Brönsted sites are dominant, with high pore volumes for the pores around 2 nm and homogeneous mesoporous distributions with low dealumination degrees are good catalysts to produce camphene [32].

The isomerization of alpha-pinene was carried out at the gas–solid interface over a range of pre-characterized bulk and supported heteropoly acid catalysts based on $H_3PW_{12}O_{40}$ in a fixed-bed flow reactor [33]. The main product of alpha-pinene isomerization is camphene, and also yielding a mixture with other monoterpenes such as limonene, terpinolenes, terpinenes, alpha-pinene, p-cymene. The catalysts included bulk and SiO_2-supported $H_3PW_{12}O_{40}$ and bulk $Cs_{2.5}H_{0.5}PW_{12}O_{40}$ possessing strong Brønsted acid sites, as well as composites prepared by supporting 15wt% $H_3PW_{12}O_{40}$ on Nb_2O_5, ZrO_2 and TiO_2 possessing both Brønsted and Lewis acid sites of moderate strength. Bulk $H_3PW_{12}O_{40}$ and $Cs_{2.5}H_{0.5}PW_{12}O_{40}$ possessing very strong Brønsted acid sites suffer from deactivation. Conversely, the catalysts prepared by supporting $H_3PW_{12}O_{40}$ on Nb_2O_5, ZrO_2 and TiO_2, although weaker acids, exhibit more stable as well as more selective performance in this reaction. The catalyst comprising $H_3PW_{12}O_{40}$ supported on TiO_2 gives a camphene yield of 51% and a total camphene and limonene yield of 58% at 200ºC after15h on stream [33].

Isomerization of alpha-pinene was carried out in presence of Ti-SBA-15 catalyst [34]. Its use enables the achievement of almost 100 mol% conversion of alpha-pinene after the reaction time of 6 h. It was shown that titanium incorporated in the silica structure significantly enhances the activity of the Ti-SBA-15 catalyst in the isomerization process, but too much titanium added to the crystallization gel results in the formation of TiO_2, which causes a decrease in the catalytic activity. This study has shown that the best titanium content in the Ti-SBA-15 catalyst is 1.1 wt% (molar ratio of silicon to titanium in the crystallization gel of 20:1). In addition, the second stage of the current research allowed us to choose the most favorable reaction conditions for the apha-pinene isomerization process, which was at a temperature of 180ºC, reaction time of 6 h, and catalyst content of 15 wt%. It is interesting and worth pointing out that the considerable elongation of the reaction time results in follow-up reactions in which monocyclic terpenes are isomerized to p-cymene. The results of the most beneficial conditions of alpha-pinene isomerization showed that the Ti-SBA-15 catalyst can be used in future industrial processes because it gives better results than the TiO_2 catalyst and at milder conditions [34].

Isomerization of α-pinene to camphene was studied over the Au/Al_2O_3 catalyst [35]. According to TPO and UV–vis-mass in situ, it was shown that deactivation is caused by adsorption of hydrocarbons on the gold species. The effects of isomerization temperature, initial α-pinene concentration and gas atmosphere on catalyst deactivation with time-on-stream were investigated experimentally and correlated with the proposed deactivation function. It was found that the increase in the α-pinene concentration in the feed mixture leads to faster catalyst deactivation. When the α-pinene concentration was controlled at a relatively low level such as 0.4 vol.%, especially in hydrogen atmosphere, the total deactivation rate was negligible. Being highly selective, the Au/Al_2O_3 catalyst undergoes strong deactivation and thus can hardly be regarded as suitable for commercial application at least in fixed bed reactors. Future modifications of the reaction conditions and/or the catalytic system might improve the performance in terms of stability enabling industrial implementation [35].

The liquid-phase isomerization of α-pinene was studied using Cr^{3+}- and Fe^{3+}-loaded clinoptilolites [36]. The major products were found to be camphene and limonene. The exchange of natural zeolite, clinoptilolite, with Fe^{3+} and Cr^{3+} ions changes the amount of surface acid sites without modifying the sample surface area, pore diameter and crystallinity. Fe^{3+}- and Cr^{3+}-loaded clinoptilolites exhibit higher Bronsted acidity compared to raw clinoptilolite. The catalytic activity strongly depends on the type and amount of the supported element. Selectivity to camphene was superior for the 0.7Fe-CL catalyst than those observed for the Raw-CL and 0.4Cr-CL catalyst. Over raw clinoptilolite, the proportion of limonene and other isomers reached ~65% while it was ~15% with 0.7Fe-CL and ~56% with 0.4Cr-CL. The results also show that the selectivity to camphene increases with increasing reaction time reaching up to about ~72% at 428 K over 0.7Fe-CL catalyst. The results of a reuse study illustrated that this mesoporous catalyst is structurally stable and provides a simple procedure for the selective synthesis of camphene which is potentially important for some industrial applications [36].

The liquid-phase isomerization of α-pinene and 3-carene were studied over acid-modified montmorillonite, kaolinite (Mongolia) and illite (Belarus) clays, as well as a commercial aluminosilicate catalyst AS-36 [37]. An increase of acid treatment temperature of natural montmorillonite, illite, and kaolinite clays resulted in removal of the framework (Al^{3+}, $Fe^{2+}/^{3+}$, Mg^{2+}) and exchangeable (K^+, Na^+, Ca^{2+}) cations, as well as an increase of the specific surface area up to 61–131 m^2/g. Modified montmorillonite clay exhibited the largest specific surface. Only weak acid sites are present on the surface of kaolin and illite clays, while in montmorillonite clays there are weak as well as medium acid sites. Commercial aluminosilicate AS-36 has a broad spectrum of sites with different (weak, medium and strong) acidity. The largest conversion of alpha-pinene (100% after 6 h) was observed in the presence of illite clay treated with 10% HCl at 90°C. The presence of only weak and medium acidic sites in the clays resulted in the formation of camphene (50.1–54.3%) as the main reaction product. Conversion of 3-carene increases with an increase in the total catalyst acidity. The reaction products were 2-carene (11.9–15.0%), mentadienes (35.5–48.5%), menthenes (7.9–17.1%) and cymenes (10,1–20.4%). The amount of mentadienes in the reaction mixture decreased while the opposite was observed for menthenes and cymenes with an increase of concentration and strength of the acid sites [37].

The vapour-phase isomerization of α-pinene was studied over Au supported on γ-Al_2O_3 [38]. The catalysts were prepared from $HAuCl_4$ as a gold precursor by an impregnation procedure including treatment of the reduced sample with NaOH for neutralization of acidic sites on the support surface. Activity significantly exceeded the catalytic performance of neat Al_2O_3 pretreated with NaOH, which indicates that gold is necessary for this kind of Wagner–Meerwein skeletal rearrangement. At low α-pinene concentration in the inlet reaction mixture α-pinene conversion over Au/γ-Al_2O_3 catalyst is not significantly changed during the catalytic run, at least, for 7–8h. However, increasing the alpha-pinene concentration in the inlet reaction mixture induces the intense deactivation of the catalyst. HRTEM and TPO results show that the deposition of carbonaceous species, which may block the active sites of the Au/γ-Al_2O_3 catalyst, is the main reason for the catalyst deactivation. The calcination of the deactivated catalyst in flowing O_2 at temperatures up to 923 K restores its activity and selectivity in α-pinene isomerization to camphene almost completely. In an ongoing study, we are optimizing the composition and conditions of preparation of gold catalysts for this reaction and elucidating the reason for high activity and selectivity of supported gold in the isomerization of alpha-pinene to camphene [38].

The isomerization of alpha-pinene was studied in both gas and liquid phases over a series of faujasite-like zeolites having different framework compositions [39]. A comparison with an amorphous silicoaluminate having 13 wt% of Al_2O_3 (SA) was made. Camphene + limonene were the major products observed in the liquid phase at 393 K at conversion levels lower than 90%. At 473 K, where the reaction in the gas phase took place, limonene suffered an ulterior transformation mainly into terpinenes and terpinolenes and a faster deactivation of the zeolite catalysts as compared to the one in the liquid phase was observed. This behavior contrasts with the high stability shown by SA in the gas phase. Over the wider pore system of SA, undesired heavy products were formed in the liquid phase at the expense of the terpene isomers. However, in the gas phase this ulterior transformation of monoterpenes was minimized. A maximum activity for the zeolite catalysts was observed when the number of AlIV per unit cell reached about 17, a value where the population of Si(0Al) sites has been shown to reach the maximum [39].

The isomerization of alpha-pinene was studied over several acid-treated natural zeolite catalysts rich in clinoptilolite [40]. Zeolite samples were contacted with HCl at different concentrations at 30° C or at 60°C for 3 and 24 hours. Catalysts prepared were characterized by XRD studies, surface area, and acidity measurements. Two main groups of products were formed: bi-tricycle products (camphene and tricyclene) and monoterpenes (especially limonene). At high conversions, limonene isomerizes to other terpenes. Acid treatment decreases the selectivity to limonene. Low selectivity to limonene is likely mainly since secondary reactions of limonene proceed faster on acid-treated zeolite catalysts [40].

Alpha-Pinene isomerization was studied over zeolite beta [41]. This zeolite in small crystal and of low Si/Al ratio was successfully modified by alkaline treatment. The micro- and mesoporous hierarchical structure can be formed without distinct change in the crystallinity of zeolite beta after treatment with 0.1–0.4 M NaOH solutions. The acidic properties of zeolite beta are well preserved when using 0.1–0.2 M NaOH solutions. A concentration of NaOH solution as high as 0.7 M results in the largest mesopore volume up to 0.69 $cm^3.g^{-1}$, but coupled with evidently reduced microporosity, crystallinity and acidity. Though the best catalytic activity of α-pinene isomerization is observed on the sample treated by 0.7 M NaOH solution, which indicates that the mesopore formation plays a crucial role in this reaction. 0.2–0.4 M NaOH solutions are considered as the optimal concentration range for zeolite beta of low Si/Al ratio when aiming at high mesoporosity coupled with better preserved microporosity and acidic properties [41].

Liquid phase alpha-pinene isomerization was carried out over various aluminosilicate catalysts, MSU-S(BEA) and MSU-S(Y) mesoporous molecular sieves with different Si/Al ratios [42]. They were prepared and characterized by XRD, XRF, N_2 adsorption, ^{27}Al MAS NMR, NH_3-TPD and 2,6-di-tert-butyl-pyridine adsorption. Among them, MSU-S(BEA)-67 has a benign activity compared to other catalysts reported before due to high accessible acid sites caused by its large pore channel and hydrophobic surface. The catalyst was rather stable, and the yield drops about 10% after three regeneration processes, which was caused by slow dealumination of the zeolite-like subunits in the framework wall [42].

The isomerization reaction of alpa-pinene was carried out in the presence of sulfated zirconia (SZ) and pre-treated SZ catalyst at several temperatures [43]. The thermal treatment affects the ratio between Brönsted and Lewis sites. Results were contrasted with those obtained with SZ catalyst without thermal treatment. If only Lewis acid sites are present (zirconia, Z), there is no activity, while if only Brönsted acid sites are present (catalyzed with H_2SO_4), there is scarce activity. When both sites are present, the activity is higher. The treatment temperature has an important effect on the reaction activity, but it has no drastic effect on the selectivity to camphene. The activity and the ratio between bicyclic and monocyclic compounds present a maximum for the SZ250 catalyst. TPD of adsorbed pyridine shows that there is a decrease in the weak-medium acidity with the increase of pre-treatment temperature SZ > SZ250 > SZ350 = SZ500. Also, the strong acidity decreases in the same order, but the ratio weak-medium/strong acidity shows a maximum for SZ250 [43].

The liquid phase isomerization of alpha-pinene was studied over X and Y zeolite catalysts at 150°C [44]. The hydrothermal treatment of Y and X zeolites leads to the framework dealumination. Simultaneously, extra-framework aluminium species are formed. High degrees of dealumination can be achieved by subjecting samples of zeolite Y to temperatures up to 750°C without a significant loss of crystallinity. The increase of the dealumination degree, expressed by the decrease in NA or by the corresponding increase in NEFAL, corresponds to an increase of the Lewis acidity when compared to the Brönsted acidity. Upon framework dealumination, the extra-framework aluminium species formed are mainly penta- and hexacoordinated aluminium. Some non-framework tetrahedral aluminium seems to also be formed. The activity for a-pinene isomerization and selectivity towards camphene are strongly influenced by the numbers of framework aluminium atoms per unit cell (NA1) and of non-framework aluminium atoms per unit cell (NzFAL). On the other hand, when A1-Na is equal to zero the activity does not become equal to zero, meaning that EFAL species also play a direct role as acid sites. Also for high values of NA1, selectivity towards camphene decreases when NZFAL increases. The minimum of selectivity seems to correspond to the catalyst samples where the total number of framework aluminium atoms equals the number of extra-framework penta- and hexacoordinated aluminium atoms. For further degrees of dealumination, the selectivity strongly increases with NZFAL and even reaches the highest values for the amorphous samples HYD850 and HXD750. The catalyst deactivation is faster for the HY parent material or for the dealuminated samples prepared at the highest temperatures. Probably this deactivation is due to a-pinene olygomerisation. It can take place either inside the supercages or on the EFAL clusters formed at high temperature on the zeolite's outer surface. In any case the final effect would be the blockage of the zeolite's pore system. Selectivity increases with time on stream, probably as a direct consequence of the adsorption of increasing amounts of organic compounds. These would act as acid sites, weaker than those corresponding to the tetrahedral aluminium [44].

Cyclization

The cyclization of pseudoionone yields a mixture of alpha-ionone, beta-ionone and gamma-ionone. By careful control of reagent and reaction conditions, either the alpha- and beta- isomer can be favoured. The alpha-ionone has a violet odour and is widely used in perfumery and flavours. Beta-ionone is the main precursor of Vitamin A and beta-carotene [3, 45-47]. The liquid-phase cyclization of pseudoionone was carried out over USY zeolites [48]. The cyclization of pseudoionone yields a mixture of alpha-ionone, beta-ionone and gamma-ionone over USY zeolites with different Si/Al ratio. The initial activity increased with the Si/Al ratio of USY catalyst until a maximum, which is obtained with USY3. USY3 zeolite showed the highest catalytic activity. Good catalytic stability of USY3 was observed, after the second use [48].

The liquid-phase cyclization of pseudoionone to ionones was carried out in the presence of tungstophosphoric acid supported on silica [49]. The amount of heteropolyacid on catalysts is between 18.8 and 58.5% HPA. The pseudoionone conversion to ionones increased linearly with the Brønsted acid site density until the HPA loading approached the monolayer saturation coverage. For higher HPA contents, the pseudoionone adsorption and conversion were hampered because of spatial constraints that diminished the reactant accessibility to the proton active sites. The highest ionone yield, 79%, was obtained on a 58.5 wt% HPA/SiO_2 catalyst at 383 K and is comparable to the best values reported in literature for the homogeneously catalyzed reaction using sulfuric acid. The ionone isomer distribution was modified by varying both the temperature and the reaction time. Unsupported HPA exhibits a very poor activity for converting pseudoionone. The isomer selectivity is enhanced at high reaction temperatures and times because of the transformation of the least stable gamma-ionone in alpha-ionone. The most stable beta-isomer is not consecutively transformed into other products and contributes up to 20% of the total ionones [49].

Pseudoionone cyclization into ionones was carried out in the presence of Amberlyst 35W, $SiO_2–Al_2O_3$, unsupported and silica supported heteropolyacids (HPA/SiO_2) and zeolite HBEA [50]. It was observed that the ionone formation was preferentially promoted on solid acids containing a high density of surface Brønsted acid sites such as Amberlyst 35W and silica-supported heteropolyacids. The higher catalytic activity of HPA/SiO_2 is due not only to their strong Brönsted acidity but also to the fact that they are high surface area materials with no steric hindrance for PS cyclization. It was also observed that the ionone synthesis is more favorably carried out at high reaction temperatures, on high-loading silica-supported heteropolyacids and using toluene as a solvent. Ionone isomer distribution can be controlled by changing the reaction conditions [50].

The cyclization of pseudoionone into ionone isomers was carried out over Amberlyst 35W, silica supported triflic acid (TFAS) and silica-supported heteropolyacid (HPAS) [51]. It was observed that the catalyst activity depends on the surface acid site strength. So, the initial activity order for obtaining ionones is TFAS > Amberlyst 35W ≈ HPAS. Initially, pseudoionone is converted to ionone mixtures with an approximate statistical composition (40% alpha-ionone, 20% beta-ionone and 40% gamma-ionone), regardless of the catalyst Brønsted acid site strength [51].

The cyclization of pseudoionone into alpha-ionone was carried out over heteropolyacid supported on SBA-15 in the liquid-phase at 363 and 373 K using a batch reactor [52]. It was observed that the high catalytic activity of PW_{12}/SBA-15 materials is due to their strong Brønsted acidity, high dispersion of active phase and to the absence of steric constraints for pseudoionone cyclization. The main product is alpha-ionone with high yield (about 60% at 373 K after 2 h). It was also observed that PW_{12}/SBA-

15 is a more active catalyst in comparison with heteropolyacid supported on commercial silica [52].

The cyclization reaction of pseudoionone was carried out over solid acid catalysts of $SO_2^{4-}/TiO_2/MCM\text{-}41$ and $S_2O_2^{8-}/TiO_2/MCM\text{-}41$, which were prepared via a grafting method and sulfate/persulfate promotion [53]. Both catalysts exhibited desirable catalytic activity and better selectivity for cyclization reaction of pseudoionone compared to conventional SO_2^{4-}/TiO_2, although the grafted TiO_2 are in an amorphous form and at a low amount. For $S_2O_2^{8-}/TiO_2/MCM\text{-}41$, higher S content and more Brønsted acid sites can be achieved by persulfation, which is favorable for the protons participated cyclization reaction. [53].

Cyclization of pseudoionone to ionone isomers was carried out over a solid acid catalyst (FPSS) with two different acid functional groups: aromatic sulfonic, like that of Amberlyst, and perfluoroalkanesulfonic, like that of Nafion [54]. By changing synthetic conditions, the loading of the two acid functional groups can be modulated. It was observed that with perfluoroalkanesulfonic acid and much exchange capacity (or acid amount), the FPSS resins showed high activity, good alpha-ionone selectivity and high yield in cyclization of pseudoionone [54].

A synthesis of ionone isomers was performed in the presence of macroreticular p-(sulfonic-perfluoroalkylated)polystyrene (FPS). The influences of the type, the chain length of perfluoroalkylating reagent as well as the extent of perfluoroalkylation on the resin's properties were studied. The FPS resins exhibited higher activity, selectivity and reusability than several commercial cation-exchange resins in the synthesis of ionone [55].

The cyclization reaction of pseudoionone to ionone isomers was investigated in the presence of a series of sulfated silica MCM-41 samples denoted as SM and ASM, which were prepared by calcination at different temperatures using H_2SO_4 and $(NH_4)_2SO_4$ solutions as the promoter, respectively [56]. The catalytic performance of SM-450 and ASM-450 (the samples SM and ASM were calcined at 450°C) was comparable with that of the commercial acid resin Amberlyst-15, and the chelating structure of SO_4^{2-}/SiO_2, which was generally considered to be inactive, was responsible for the high activity. ASM-450 showed better catalytic performance compared with Amberlyst-15 at relatively low temperature, and it could be reused 5 times with good activity. The catalyst could be regenerated by calcination and sulfation, and its activity remained high although the mesostructure was destroyed during the regeneration process [56].

CONCLUSION

Terpenes are natural products available from essential oils, present on biomass. They are an important biomass-based feedstock for the flavour, fragrance and pharmaceutical industries. Traditionally, the conversion of terpenes is carried out in the presence of homogenous catalysts, which have some disadvantages, such as production of toxic waste and its difficulty in separation from reaction mixture. To transform these different processes into "green process", these reactions have been carried out over different heterogeneous catalysts. The use of heterogeneous catalysts for terpenes conversion has been applied in different reactions.

REFERENCES

[1] Mäki-Arvela, P; Holmbom, B; Salmi, T; Murzin, DY. *Cataly. Rev.*, 2007, 49, 197-340.

[2] Gusevskaya, EV. *ChemCatChem*, 2014, 6, 1506–1515.

[3] Corma, A; Iborra, S; Velty, A. *Chem. Rev.*, 2007, 107, 2411-2502.

[4] Gupta, P; Paul, S. *Catal. Today*, 2014, 236, 153-170.

[5] Hensen, K; Mahaim, C; Hölderich, WF. *Appl. Catal. A: Gen.*, 1997, 149, 311-329.

[6] Castanheiro, JE; Guerreiro, L; Ramos, AM; Fonseca, IM; Vital, *J. Stud. Surf. Sci. Catal.*, 2008, 174, 1319-1322.

[7] Pito, DS; Fonseca, IM; Ramos, AM; Vital, J; Castanheiro, JE. *Chem. Eng. J.*, 2009, 147, 302-309.

[8] Pito, DC; Matos, I; Fonseca, IM; Ramos, AM; Vital, J; Castanheiro, JE. *Appl. Catal. A: Gen.*, 2010, 373, 140–146.

[9] Caiado, M; Pito, DS; Castanheiro, JE. in *Environmentally Benign Catalysts: For Clean Organic Reactions*, Patel, A.; Ed.; Springer, 2013, 153- 164.

[10] Matos, I; Silva, MF; Ruiz-Rosas, R; Vital, J; Rodríguez-Mirasol, J; Cordero, T; Castanheiro, JE; Fonseca, IM. *Microp. Mesop. Mater.*, 2014, 199, 66–73.

[11] Castanheiro, JE; Fonseca, IM; Ramos, AM; Vital, J. *Microp. Mesop. Mater.*, 2017, 249, 16-24.

[12] Caiado, M; Machado, A; Santos, RN; Matos, I; Fonseca, IM; Ramos, AM; Vital, J; Valente, AA; Castanheiro, JE. *Appl. Catal. A: Gen.*, 2013, 451, 36- 42.

[13] de Meireles, ALP; Costa, MS; Rocha, KAS; Kozhevnikova, EF; Kozhevnikov, IV; Gusevskaya, EV. *Chem Cat Chem*, 2014, 6, 2706-2711.

[14] Kozhevnikov, IV; Sinnema, A; van der Weerdt, AJA; van Bekkum, H. *J. Mol. Catal. A: Chem.*, 1997, 120, 63-70.

[15] Botella, P; Corma, A; López-Nieto, JM; Valencia, S; Lucas, ME; Sergio, M. *Appl. Catal. A: Gen.*, 2000, 203, 251–258.

[16] Gainsford, GJ; Hosie, CF; Weston, RJ. *Appl. Catal. A: Gen.*, 2001, 209, 269–277.

[17] Yadav, MKr; Patil, MV; Jasra, RV. *J. Mol. Catal. A: Chem.*, 2009, 297, 101-109.

[18] Machado, J; Castanheiro, JE; Matos, I; Ramos, AM; Vital, J; Fonseca, IM. *Microp. Mesop. Mat.*, 2012, 163, 237–242.

[19] Robles-Dutenhefner, PA; da Silva, KA; Rafi, M; Siddiqui, H; Kozhevnikov, IV; Gusevskaya, EV. *J. Mol. Catal. A: Chem.*, 2001, 175, 33–42.

[20] da Silva, KA; Kozhevnikov, IV; Gusevskaya, EV. *J. Mol. Catal. A: Chem.*, 2003, 192, 129-134.

[21] de Meireles, ALP; da Silva-Rocha, KA; Kozhevnikov, IV; Gusevskaya, EV. *Appl. Catal. A: Gen.*, 2001, 409–410, 82-86.

[22] Castanheiro, JE; Fonseca, IM; Ramos, AM; Vital, J. *Catal. Commun.*, 2008, 9, 2205-2208.

[23] Valente, H; Vital, J. *Stud. Surf. Sci. Catal.*, 1997, 108, 555-562.

[24] Vital, J; Ramos, AM; Silva, IF; Valente, H; Castanheiro, JE. *Catal. Today*, 2000, 56, 167-172.

[25] Vital, J; Ramos, AM; Silva, IF; Castanheiro, JE. *Catal. Today*, 2001, 67, 217–223.

[26] Castanheiro, JE; Fonseca, IM; Ramos, AM; Vital, J. *Catal. Today* 2003, 82, 187–193.

[27] Castanheiro, JE; Fonseca, IM; Ramos, AM; Oliveira, R; Vital, J. *Catal. Today*, 2005, 104, 296–304.

[28] Comelli, N; Avila, MC; Volzone, C; Ponzi, M. *Cent. Eur. J. Chem.*, 2013, 689-697.

[29] Yang, G; Liu, Y; Zhou, Z; Zhang, Z. *Chem. Eng. J.*, 2011, 168, 351-358.

[30] Ávila, MC; Comelli, NA; Rodríguez-Castellón, E; Jiménez-López, A; Flores, RC; Ponzi, EN; Ponzi, MI. *J. Mol. Catal. A: Chem.*, 2010, 322, 106-112.

[31] Volzone, C; Masini, O; Comelli, NA; Grzona, LM; Ponzi, EN; Ponzi, MI. *Appl. Catal. A: Gen.*, 2001, 214, 213–218.

[32] Besün, N; Özkan, F; Gündüz, G. *Appl. Catal. A: Gen.*, 2002, 224, 285–297.

[33] Alsalme, A; Kozhevnikova, EF; Kozhevnikov, IV. *Appl. Catal. A: Gen.*, 2010, 390, 219–224.

[34] Wróblewska, A; Sreńscek-Nazzal, PMJ; Sadłowski, M; Koren, ZC; Michalkiewicz, B. *Microp. Mesop. Mat.*, 2018, 258, 72-82.

[35] Solkina, YS; Reshetnikov, SI; Estrada, M; Simakov, A; Murzin, DY; Simakova, IL. *Chem. Eng. J.*, 2011, 176–177, 42-48.

[36] Akgül, M; Özyağcı, B; Karabakan, A. *J. Ind. Eng. Chem.*, 2013, 19, 240-249.

[37] Sidorenko, AY; Aho, A; Ganbaatar, J; Batsuren, D; Utenkova, DB; Sen'kov, GM; Wärnå, J; Murzin, DY; Agabekov, VE. *Mol. Catal.*, 2017, 443, 193-202.

[38] Simakova, IL; Solkina, YS; Moroza, BL; Simakova, OA; Reshetnikov, SI; Prosvirin, IP; Bukhtiyarov, VI; Parmon, VN; Murzin, DY. *Appl. Catal. A: Gen.*, 2010, 385, 136-143.
[39] López, CM; Machado, FJ; Rodríquez, K; Arias, D; Méndez, B; Hasegawa, M. *Catal. Lett.*, 1999, 62, 221–226.
[40] Ünveren, E; Gündüz, G; Cakicioğlu–Özkan, F. *Chem. Eng. Comm.*, 2005, 192, 386–404.
[41] Wu, Y; Tian, F; Liu, J; Song, D; Jia, C; Chen, Y. *Microp. Mesop. Mat.*, 2012, 162, 168-174.
[42] Wang, J; Hua, W; Yue, Y; Gao, Z. *Bioresour. Techn.*, 2010, 7224-7230.
[43] Comelli, NA; Ponzi, EN; Ponzi, MI. *Chem. Eng. J.*, 2006, 117, 93-99.
[44] Severino, A; Esculcas, A; Rocha, J; Vital, J; Lobo, LS. *Appl. Catal. A: Gen.*, 1996, 142, 255-278.
[45] Musa, S; Braun, M. *Ullman's Encyclopedia of Industrial Chemistry*, sixth ed., Plenum Press, New York, 2002, p. 540.
[46] Brenna, E; Fuganti, C; Serra, S; Kraft, P. *Europ. J. Organ. Chem.*, 2002, 967-978.
[47] Bedoukian, PZ. In *Perfumery and Flavoring Synthetics*, Elsevier, New York, 1967, p. 99.
[48] Caetano, CS; Caiado, M; Castanheiro, JE. In *Heterogeneous Catalysts: Design, Applications and Research Insights*; Jensen, K. Ed.; Nova Science Pub. Inc., New York, pp. 177-193.
[49] Díez, VK; Marcos, BJ; Apesteguía, CR; Di Cosimo, JI. *Appl. Catal. A: Gen.*, 2009, 358, 95–102.
[50] Díez, VK; Apesteguía, CR; Di Cosimo, JI. *Cataly. Lett.*, 2008, 123, 213–219.
[51] Díez, VK; Apesteguía, CR; Di Cosimo, JI. *Catal. Today*, 2010, 149, 267–274.
[52] Rachwalik, R; Michorczyk, P; Ogonowski, J. *Catal. Lett.*, 2011, 141, 1384–1390.
[53] Guo, Dai-Shi; Ma, Zi-Feng; Jiang, Qi-Zhong; Xu, Hai-Hong; Ma, Zheng-Fei; Ye, Wei-Dong. *Cataly. Lett.*, 2006, 107, 155-159.

[54] Lin, Z; Ni, H; Du, H; Zhao, C. *Catal. Commun.*, 2007, 8, 31–35.
[55] Lin, ZH; Guan, CJ; Feng, XL; Zhao, CX. *J. Mol. Catal. A: Chem.*, 2006, 247, 19–26.
[56] Haihong, X; Daishi, G; Qizhong, J; Zifeng, M; Wanjun, L; Zheng, W. *Chinese J. Catal.*, 2006, 27, 1080–1086.

In: Terpenes
Editor: Albin Bjarke

ISBN: 978-1-53613-622-7

Chapter 3

VOLATILE TERPENE EMISSIONS FROM WOOD PROCESSING IN PULP MILLS

Frédérique Bertaud*[*]*, PhD
InTechFibres Team, Centre Technique du Papier
Grenoble, France

ABSTRACT

Turpentine, which is composed of the volatile terpenes fraction from conifer wood, is the most widely produced essential oil. Of the 330 000 T of turpentine produced worldwide every year, more than half comes from kraft pulp mills. This sulfate turpentine is recovered at a rate of 2-15 kg/T of air-dried pulp, and is composed mainly of 45-95% α-pinene and 0.5-28% β-pinene [3] [5]. The monoterpene recovery rate and composition depend essentially on the wood species used for wood-pulp manufacturing (high terpene content in pine species) [9], but also on the wood processing operations prior to pulping [1] [3]. Avoiding long storage times of wood raw materials prior to pulping will lead to a higher turpentine yield. An exhaustive process-line assessment could help a kraft pulp mill improve its recovery yield [25]. Highly volatile terpene emissions also occur during thermal/mechanical processes in sawmills and thermomechanical pulp (TMP) mills. Between 0.4-1 kg of turpentine

* Corresponding Author Email: Frederique.Bertaud@webctp.com.

per ton of TMP could be recovered in a TMP mill [25] [27], thereby reducing volatile organic compound (VOC) emissions accordingly. Like the gum turpentine obtained from tree tapping, TMP tupentine can be marketed at a price four to five times higher than that of sulfate turpentine from kraft pulp contaminated with malodorous sulfurous compounds.

Keywords: terpenes, volatile, VOC, turpentine, sulfate turpentine, wood, wood extractives, wood pulp, chemical pulping, mechanical pulping, recovery, analysis, process diagnostics

1. Introduction

Volatile terpenes are natural wood extractive compounds, with amounts varying from 0.02 to 2% [1]. Monoterpenes (α-pinene, β-pinene, carene, limonene, etc.) are highly volatile and contribute to the odor of wood. In wood-based industries, volatile terpenes are released especially during debarking, chipping, steaming, mechanical and chemical pulping of softwoods. In the kraft pulping process, which produces amost 70% of the world wood-pulp production (i.e., virgin pulp) [2], volatile monoterpenes are mainly in the form of sulfated turpentine (i.e., sulfated pine resin), which represents up to 15 kg/T of pulps [1]. Taking into account both the recovery and marketing of tall oil, wood extractives from black liquor represent 1-1.5% or even more of the kraft mill's total revenues [3].

Most terpenes, used in the flavoring and fragrance industry, are derived from turpentine: 50% from the volatile fraction of kraft pulping liquors (Europe and North America) and 50% from pine tree tapping (China). Turpentine gum is becoming scarce with prices increasing drastically (up to € 2,500/T today) [4], and wood-kraft pulp mills are becoming the main terpene producers [5]. Turpentine is by far the most widely produced essential oil in the world (330 000 tons per year) and consequently the privileged source for applications requiring large-scale supplies [5] [6]. The development of green chemistry has led to terpenes being in even greater demand, especially for use as renewable solvent and platform chemicals [5] [9]. Often treated as just a by-product of pulp processing, turpentine is still underestimated by most kraft pulp mills with almost no dedicated management of wood raw materials, only a few process line diagnostics and rarely any final turpentine quality check. In wood thermal/mechanical processes and especially thermomechanical pulping (TMP), volatile terpenes are released mainly during the 1st and 2nd refining stages, and contribute more than half of the total VOCs in air emissions [7]. Only one TMP mill in Europe is currently recovering turpentine from TMP steam rejects [8].

2. Terpene Behavior during Wood Pulp Processing

2.1. Wood Composition, Ultra-Structure and Terpene Contents

Apart from its main structural polymeric constituents (i.e., cellulose, hemicellulose and lignins), wood also contains extractable components, known as wood extractives, a term that covers a wide variety of low molecular mass components, which play a major protective role against fungi and insects. Lipophilic (hydrophobic) wood extractives, also known as wood resins, are extractable with non-polar solvent and are composed of terpenes and terpenoids (including volatile terpenes, resin acids and steroids), esters of fatty acids (fats and waxes), fatty acids and fatty alcohols and alkanoids [9] [10] [11].

Depending on their principal functions, wood cells can be divided into different groups: i) cells performing a liquid transport function, ii) cells providing the necessary mechanical strength to the tree, and iii) cells acting as a means of storing reserve nutrient supplies [11]. In softwood (i.e., coniferous), tracheids (2-6 mm long fibers making up 90-95% of wood cells) perform both conduction and structural strength functions. The majority of fibers forming wood trunks are dead cells. The storage cells are thin-walled parenchyma cells that transport and store nutrients, predominantly arranged in radially oriented rays. Softwood tissues also possess vertical and horizontal resin canals. In general, resin canals are tube-like, intercellular cavities lined with the epithelial parenchyma cells, which secrete oleoresin into the resin canals [10] [11]. Resin canals are natural and constant features of pine (*Pinus* spp.), spruce (*Picea* spp.), larch (*Larix* spp.), and Douglas fir (*Pseudotsuga menziesii*) species. Other conifers, such as fir (*Abies* spp.), cedar (*Cedrus* spp.), tsuga and sequoia, form only traumatic resin canals, that are generally axial, and some resin pockets in response to wounding. In tropical hardwood, resin is also found in other secretory tissues, e.g., in oil cells and latex cells [10].

The main function of canal resin is to seal wounds mechanically with an oleophilic material, resistant to most fungi. Canal resin is generally a mixture of cyclic terpenes and terpenoids. About 25-30% of monoterpenes and 1% of sesquiterpenes are volatile, and the rest are non-volatile diterpenoic compounds (mainly resin acids with a small amount of diterpene alcohols). In contrast to resin canals, parenchyma cells are mainly composed of fats and steryl esters of fatty acids, and can contain alkanol esters and hydroxy fatty acids [10].

Of the fourteen volatile monoterpenes that occur in the oleoresin of many conifers (Table 1), the most frequently reported volatile terpenes are α-pinene, β-pinene and limonene, but 3-carene, camphene, myrcene and β-phellandrene are also widespread (Table 2). Conifer sesquiterpenes are also volatile and humulenes, caryophyllenes and longifolene are the most extensively reported types. Some terpenoids (ethers, alcohols, aldehydes, esters and carboxylic derivatives) can also occur but are less volatile [1].

The volatile monoterpene content in conifer wood varies with species, ranging from 0.02% to 2%. It also varies among individual trees depending on genetic and environmental factors [1] [9]. In a same tree, the monoterpene composition of the wood will also vary, as there is a greater proportion of resin canals in the lower part of the stem and in the crown than in the main part of the stem. In pine wood, the canals are concentrated in the heartwood (inner part of the tree) and root, whereas in spruce they are more evenly distributed throughout the whole wood [1]. Pine wood contains more and larger resin canals than spruce wood, and pine heartwood can have a resin content that is two or three times higher than that of sapwood [10] [11] [12].

Table 1. Chemical structures of some common monoterpenes commonly found in spruce and pine [1]

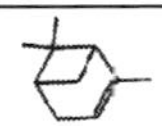	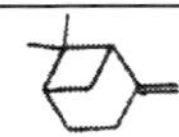		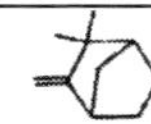	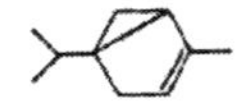	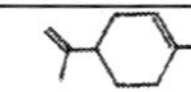	
α-pinene	β-pinene	3-carene	camphene	α-thujene	limonene	β-phellandrene
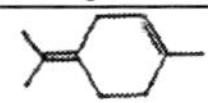	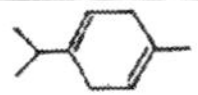		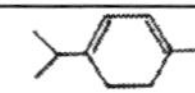	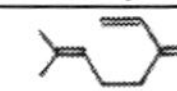	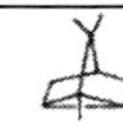	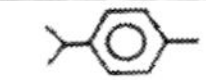
terpinolene	γ-terpinene	α-phellandrene	α-terpinene	myrcene	tricyclene	p-cymene

Table 2. Volatile monoterpene content in some conifer woods (*industrial processing wood)

Composition (%, w/w)	Norway spruce (*Picea abies*)			Scots pine (*Pinus sylvestris*)	Maritime pine (*Pinus pinaster*)	Eastern white pine (*Pinus strobus*)	Fir (*Abies alba*)	
	[1]*	[9]	[13]*	[1]*	[13]*	[9]	[9]	[13]*
α-pinene	38	58	65.2	42	92.8	67	39	70
β-pinene	28	24	26.1	3.2	2.3	18	3.0	20
3-carene	4.6	2.1	< 0.1	46	< 0.1		4.5	< 0.1
camphene	5.6	3.1	-	1.1	-	*2.9*	*< 1.0*	-
α-thujene	0.1	-	-	0.2	-	-	-	-
limonene	8.0	4.5	-	1.7	0.25	0.9	53.5	< 0.1
β-phellandrene	8.3	-	-	1.4	-	0.5	-	-
terpinolene	0.4	0.4	-	1.7	-	0.5	-	-
γ-terpinene	0.2	-	-	0.4	-	-	-	-
α-phellandrene	0.2	-	-	0.1	-	-	-	-
α-terpinene	0.2	-	-	0.1	-	-	-	-
myrcene	6.0	-	-	2	-	*0.9*	< 1.0	-
p-cymene	0.5	0.5	< 0.1	0.3	0.25	-	-	< 0.1

2.2. Volatile Terpene Emissions from Forestry and Wood Processing

Volatile monoterpene emissions occur naturally from trees and other plants although much greater emissions occur during wood processing in the forestry and wood industries. Monoterpenes easily evaporate into the air due to the relatively weak hydrophobic forces between the molecules. They are volatile and contribute to the characteristic odor of conifers. The boiling points are in the range 151°C to 186°C, and their vapor pressure at ambient temperature can be high (Table 3) [1].

The amount and composition of wood resin (including volatile monoterpenes) can change considerably during harvesting, transportation, storage and deliberate seasoning of the wood [12]. Anthropogenic emissions of these biogenic VOCs occur due to machining, logging and chipping, with terpene concentration levels up to 1000 times higher than background levels (i.e., from 100-500 $\mu g/m^3$ to 1.0-1.5 mg/m^3 compared to a background level of ~1 $\mu g/m^3$) [14] [15]. Drying, debarking, sawing and pulping of wood leads to even more volatile terpene emissions, with 50 to 550 mg/m^3 in the air of a sawmill, and emissions from the drying of sawdust chips and lumber range from 120 to 2020 mg/kg o.d. [15].

Table 3. Volatile wood terpene characteristics [1]

	Boiling point (°C)	Vapor pressure (mm Hg at 50°C)	Density (kg/l)
α-pinene	156	19	0.858
camphene	158	12	0.842
β-pinene	163	15	0.865
3-carene	170	8.8	0.844
β-phellandrene	172	-	0.846
limonene	176	8.6	0.842
p-cymene	177	7.3	0.860

In pulp mills, during storage of roundwood, terpenes are slowly released into the air. Chipping and long storage of wood chips increase terpene losses markedly. More than half of the terpene content may be lost during chip pile storage for few weeks; the losses occur mainly by evaporation into the air [14] [1]. Because of the fast release of terpenes from chips, especially in the summer season, chip storage is probably the largest source of terpene emissions to the air from pulp mills [1].

The volatility, flammability, strong aroma and solvent properties of volatile terpenes dictate that their recovery (removal) from gaseous and liquid waste streams is desirable, and in some cases, necessary in order to comply with occupational health and environmental safety regulations.

2.3. Terpenes from Chemical Pulping

Of the 184 Mt of wood pulp produced annually throughout the world, more than 77% is chemical pulp, 97% of which produced by the kraft process (127 Mt/y) [2] [16].

In the conventional kraft pulping process, a strong alkaline liquor consisting of sodium hydroxide ($NaOH$, 15-20 g/100g of o.d. wood chips, 80-120 g/l) and sodium sulfide (Na_2S, 5-10 g/100g of o.d. wood chips, 30-40 g/l) (i.e., cooking liquor or white liquor) delignifies wood chips, at 160°C-170°C for 2 to 5 hours, in a batch or continuous digester [2]. After cooking (or digestion), the spent cooking liquor (i.e., black liquor) is separated from the pulp containing the liberated fibers. The black liquor is then concentrated to 65-80% solid content by mutiple-effect evaporators and then burnt in the boiler used for recovering cooking chemicals and generating energy [17].

During kraft pulping, the volatile terpenes are volatilized from wood during the steaming of chips before cooking, and in digester degassing in the early phases of the kraft process [18]. In principle, the terpenes pass through the kraft process chemically unchanged [1]. In processes that do not apply steaming, such as displacement batch cooking, volatile terpenes are recovered from off-gases produced in warm and hot liquor accumulators. The volatile terpenes are recovered by condensation of the released gases from cooking [18]. Monoterpenes have a low solubility in water (for example, the solubility of α-pinene and p-cymene is 22 ppm and 15 ppm, respectively at 25°C [1]), and are less dense than water (Table 3). They are easily separated from condensed process water as a light oil composed of a mixture of terpenes with a small amount of a sulfurous component, the so-called 'sulfated crude turpentine' [17]. Due to their high volatility (Table 3), attention should be paid to keeping a low concentration of terpenes in the processes for safety reasons given the explosion limit for terpenes which can be as low as 0.7% [1].

The average yield of crude turpentine from pine species is 2-15 kg/t of pulp, this value being lower for spruce species (1-2 to 3 kg/t pulp) [1] [18]. The yield is highly dependent on the quality of the wood raw materials (sawmill chips, round wood chips), the wood storage time and the steaming time and temperature. Highest recovery is usually obtained from a batch process. The turpentine loss can be dramatic if the storage time exceeds several weeks, as previously mentioned (cf. § 2.2). Processing fresh wood will result in higher turpentine yields. For pine species, the proportion of chips from sawmills also affects the terpene quantities released, due to the larger amount of sapwood in sawmill chips, which contain less terpenes than heartwood [1].

Generally, softwood kraft turpentine is composed of 60-80% of α-pinene, 2-7% of β-pinene, 11-28% of 3-carene and 2-6% of other terpenes [18]. The purity of turpentine varies, depending on its steaming phase prior to cooking. Steaming produces the purest fraction, without sulfurous compounds. Crude tupentine is purified by vacuum distillation in a special distillation plant where impurities such as methyl-mercaptan (or methanethiol) and dimethyl sulfide are removed. The main fractions are i) monoterpene fractions consisting of α-pinene (50-80%), β-pinene and 3-carene and ii) a pine oil fraction consisting of hydroxylated monoterpenes [17] [19]. There are only two large sulfate turpentine distillation plants in operation in Europe, one in Finland (Arizona Chemical, Oulu), and one in France (DRT, Dax), but there are several plants in the United States [19].

In addition to the kraft process, another chemical pulping process, based on acidic-sulfite chemical treatment, represents only a small part of total pulp production (4 Mt/y) [16]. Applied to conifers, the monoterpenes released from the wood are converted by acid rearrangements and dehydration with the formation of p-cymene. The p-cymene oil can be recovered, a process that is practiced in Sweden. As sulfite cooking involves processing mainly spruce, turpentine yield is lower than in the kraft process. In addition, a large part of the wood monoterpenes is already lost during the necessary chip seasoning stage [1].

2.4. Terpenes from Mechanical Pulping

Mechanical wood pulp is obtained by grinding or milling coniferous or non-coniferous rounds, quarters, billets, etc., into relatively short fibers or by refining coniferous or non-coniferous chips. This pulp is used mainly in newsprint and wood-containing papers, such as LWC (light-weight coated) and SC (super-calendered) papers and in paperboard. Thermomechanical pulp (TMP) is produced by a mechanical process in which wood chips are softened by pre-heating under pressure prior to pressurized refining stages. Chemi-thermomechanical pulp (CTMP) is produced in a similar way to TMP, but the wood chips are subjected to preliminary chemical treatment before entering the refiner. This pulp has properties suitable for tissue, printing paper and multiply-board manufacture [20].

In mechanical pulping, volatile terpene emissions occur mainly during steaming and refining of the wood chips. The dual effect of delamination of wood fibers coupled with steam extraction leads to considerable terpene emissions [21] [22]. Terpenes are still vented to air after pre-heating wet chips at TMP mills. The most volatile terpene emissions occur during the 1st and 2nd refining stages [23] [24] [25] and account for more than half of the total VOC emissions of the TMP mill (i.e., 156 mg of volatile terpene per kg of o.d. radiata pine wood chips) [26].

Fresh wood is normally processed directly, meaning that terpene losses during storage are negligible. Volatile terpenes are rarely recovered. Only a few plants operate a turpentine recovery system, such as SCA Forest Products Ortviken mill in Sweden [27]. 0.4 kg of turpentine per ton of pulp is then collected at the mill. The mill processes mainly spruce wood, providing 320 tons of TMP-turpentine each year [27], whereas up to 1 kg/T of pulp would be possible depending on the spruce composition [1]. Turpentine recovery is integrated in the heat recovery system of the TMP mill [28]. Modern TMP heating and turpentine recovery systems are marketed by Andritz, a pulp and paper equipment supplier [29] [30].

The tupentine from the TMP process has the great advantage of not containing malodorous sulfur contaminants, unlike kraft crude sulfate turpentine [1]. In addition to providing better economic value, close to that of gum turpentine obtained from tapping trees (up to 2500 US $/t) [4] [33], turpentine recovery in a TMP mill leads to a reduction in VOC emissions to the air, which is of real importance from an environmental point of view.

Terpene emissions from the production of CTMP may be similar in composition to those from sulfate or sulfite mills, depending on the wood material and the pre-treatment chemicals [1].

3. TURPENTINE

Turpentine is the common term given to the volatile fraction isolated from pine resin. Depending on the way pine resin is isolated from the wood, three products are distinguished: i) gum turpentine obtained from tapping living trees (oleoresin); 2) sulfate turpentine recovered during kraft pulping, and 3) to a lesser extent, the exudate from wood solvent extraction. Turpentine, is by far the most widely produced essential oil in the world [5].

The turpentine world market consists of 55% oleoresin and 45% sulfated turpentine. World production is estimated to be 350 000 t/y in 2016, with China being the main producer (74%) [5] [4] [6] [33]. The turpentine price is highly dependent on the product's origin, with gum turpentine being five times more expensive than sulfate turpentine [31]. Turpentine prices are subject to negotiation although they are highly dependent on the quality and composition of the turpentine: the greater the proportion of β-pinene compared to α-pinene, the higher its value. Some compounds lower the value, such as 3-carene; this is the case for Indian gum turpentine [6] [4].

Turpentine composition depends to a large extent on the tree species and age, geographic location, as previously indicated for the volatile terpene composition of wood (§ 2.1), and also on the overall procedure used to isolate the compound. In general, however, the major components are basically monoterpenes: α-pinene (45-97%) and β-pinene (0.5-28%) with smaller amounts of other monoterpenes [4] [5] [6].

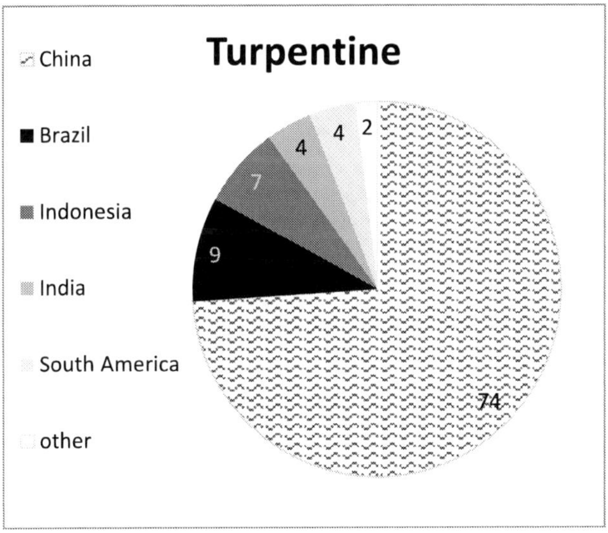

Figure 1. Main turpentine-producing countries [4].

Turpentine was mainly used as a solvent in the 1940s, and only a small part was exploited as a raw material in the chemical industry. Because of the progressive exploitation of the particular reactivity of its terpenic components, its use was completely revitalized in the 1980s with turpentine mainly being employed as Q reagent in the chemical industry.

Historically, the most common uses of turpentine components as reagents include: i) insecticides, although this application was progressively banned from the 1970s (i.e., Toxaphene from camphene), ii) synthetic pine oil production obtained by the hydration of α-pinene with aqueous mineral acid, leading mainly to α-terpineol, and used in household cleaning and disinfectant products, and iii) flavorings and fragrances and other fine chemicals such as the production of rose-alcohols from α-pinene, as well as polyterpene resins (e.g., tackifying resin polyPINs used in chewing gum and food packaging) [5]. As most of the chemicals and products derived from turpentine can also be produced from petroleum-based products, the use and application of terpenes is in direct competition with petrol-derived products.

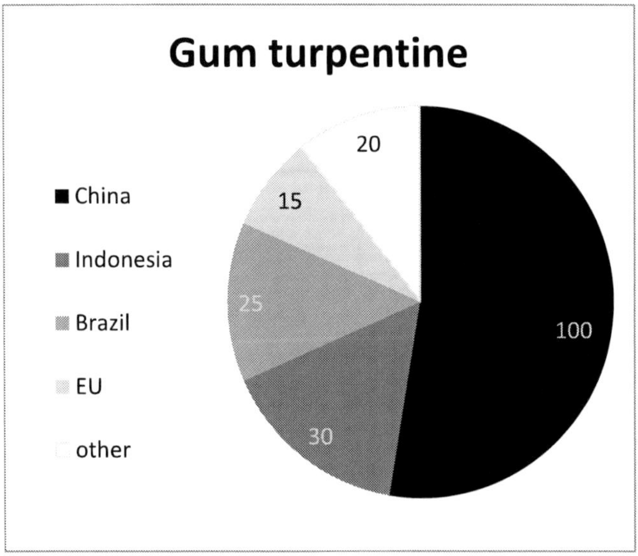

Figure 2. Main uses of turpentine [4].

3.1. Gum Turpentine

Less than 50% of the total turpentine production consists of crude oleoresin produced from tapping trees, with a continuously decreasing production. This ancestral industry is still alive in China, Brazil and Indonesia, with more than 50% of the production coming form China (Figure 3). 2-3 kg of crude oleoresin are produced per tree each year, with the same tree being exploited over a period of 20 years. One person can harvest 1500 trees/year [4] [32].

Crude oleoresin from tapping generally comprises ~20% of volatiles and ~70% of resin acids. After distillation, the crude oleoresin gives a gum turpentine rich in monoterpenes and gum rosin (rich in resin acids). The gum turpentine price has increased substantially, for example, a 300-fold increase in ten years, especially as a result of a drop in production (Chinese production has been decreasing since 2011) and a higher demand. The gum turpentine price has varied between 1500 and 2500 US$/MT over the past 7 years, with the highest values recorded in 2012-2014 [33]. The gum turpentine price is highly dependent on oil market variations and the quality of the turpentine [4] [31]. A total pinene content of 90% or greater would be regarded as good, becoming excellent as the β-pinene contribution increases to 30-40% or more. Portuguese, American and Brazilian turpentines are all high in pinenes [32]. If the pinene content drops below 70 to 80wt%, then it has limited value for derivative manufacture [6]. Its value is lowered by the presence of certain compounds, such as 3-carene, which may make up 50% or more of Indian turpentine. This product finds little use other than as a solvent. α-pinene, β-pinene and d-limonene are the important ingredients for producing different compounds such as anethol, citral, camphene, linalool, etc. [5] [6].

Pine trees are the main source of oleoresin which is produced mainly in China, Brazil, Indonesia, and USA from specific local pine species (Table 4). American and Brazilian gum turpentine contains slightly more β-pinene than that from Europea and Russia (Brazilian α/β of 60/40). In North Amercia and Europe, however most turpentine requirements consist of sulfate turpentine, but gum turpentine is also imported for fractionation and conversion into derivatives [32].

TMP turpentine composition is very similar to that of gum turpentine, as it is exempt from sulfurous contaminants. The TMP turpentine produced in the SCA Forest Products Ortviken TMP mill is indicated in Table 5: Volatile terpene content of TMP turpentine from the SCA Forest Products Ortviken mill, obtained from Norway spruce (Picea abies) [27].

3.2. Sulfate Turpentine

After distillation, sulfated turpentine (2-15 kg/t of pulp) and tall oil (30-50 kg/t) are produced from kraft black liquor. The world production of sulfated turpentine was 170 000 Mt in 2015 with North America and Europe being the main producers (Figure 4) [33]. In Europe, the main producer countries are Sweden and Finland. The French production is very small with 7000 – 8000 t/year [4]. Based on the announced expansion of softwood pulp capacity, the crude sulfate turpentine capacity will increase to about 10 000 MT/year over the next 3-5 years [33].

The market price of gum-turpentine was 500 to 700 US$/MT in Europe, whereas in the USA, it reached as high as 1400 US$/MT in 2012-2013. Over the past 3 years, the market price has continuously increased, with a 15% rise in Europe and up to 30% growth in the USA [33].

Table 4. Main oleoresin sources [4] [32]

Pinus Elliotti (Slash pine)	USA, Brazil, South Africa, Zimbabwe, Kenya
Pinus pnaster (Maritime pine)	Portugal, France
Pinus Massoniana (Masson pine)	China
Pinus Merkusii (Merkus pine)	Indonesia, Thailand
Pinus caribaea (Caribbean pine)	South Africa, Kenya
Pinus Roxburghii – syn. P. longifolia (Chir pine)	India, Pakistan
Pinus oocarpa	Mexico, Honduras
Pinus sylvestris L. (Scots pine)	Russia, Scandinavia
Pinus radiata D. (Monterey pine)	Kenya
Pinus Halepensis (Aleppo pine)	Greece
Pinus brutia	Turkey

Generally, softwood kraft turpentine is composed of 50-80% of α-pinene, 2-35% of β-pinene, 2-30% of 3-carene and 1-10% of other terpenes (Table 6) [4] [35]. In the USA, the monoterpene content of gum turpentine is very similar to that of kraft turpentine, with mainly α-pinene and β-pinene (Table 5). In France, maritime pine turpentine contains 63% of α-pinene and 26.5% of β-pinene, with small amounts of limonene, caryophyllene, longifolene and pinol [36].

Table 5. Volatile terpene contents of TMP turpentine from the SCA Forest Products Ortviken mill, obtained from Norway spruce (Picea abies) [34] [27]

Ingredient	Content (%, w/w) [27]	Distribution (%)
α-pinene	42.3	54.3 – 53.6
β-pinene	20.8	25.3 - 26.3
3-carene	3.0	3.8 - 7.1
camphene	-	1.1
limonene	10.0	7.9 – 12.7
other	2.8	3.6 - 3.8
Total	**78.9%**	**~100%**

The purity of turpentine varies depending on the kraft process steaming phase from which it is derived. Steaming produces the purest fraction, withouth sulfurous compounds. A methyl-mercaptan content of 5% was measured in the kraft turpentine produced by the SCA Forest Products Östrand mill in Sweden [27].

Table 6. Kraft turpentine composition (%, w/w) [4]

	Finland	Sweden	USA	Russia
α-pinene	55-70	50-70	40-70	55-70
β pinene	2-6	4-10	15-35	1-5
camphene	1	1	1-2	1-8
3-carene	7-30	15-40	2-10	10-25
limonene	4	1-3	5-10	3-8

Table 7. Terpene contents of USA kraft sulfated turpentine and gum turpentine [4]

Ingredient	Sulfated turpentine (%)	Gum turpentine (%)
α-pinene	65.5	65.6
β-pinene	20.4	28.1
3-carene	1.8	-
cymene	0.4	-
other	11.9	6.3

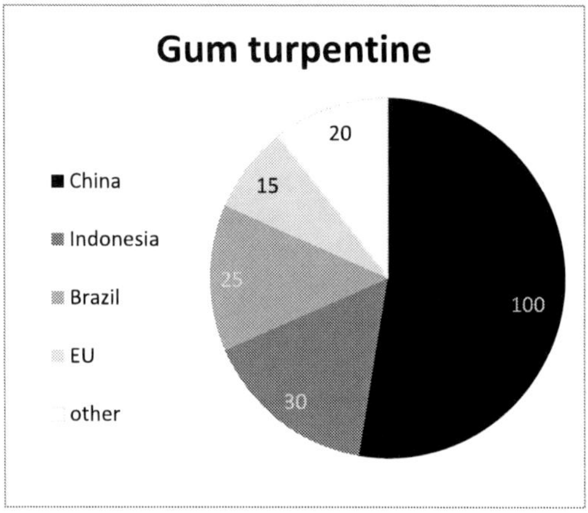

Figure 3. Main gum turpentine-producing countries.

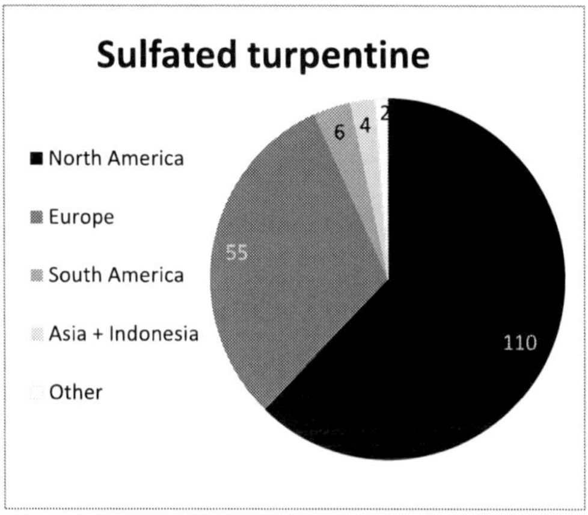

Figure 4. Main sulfated turpentine-producing countries.

4. Inventory of Volatile Terpene Emissions in Pulp Mills

In order to evaluate the emissions of volatile terpene contained in side streams of pulp mills, and therefore to identify the main source of VOC in a mill, an inventory of emissions can be established by identifying and quantifying the main volatile terpenes in aqueous, gaseous and/or wood/fiber samples collected along the process lines [23] [25] [37].

In a recent study, volatile terpenes emissions were exhaustively mapped in a kraft pulp mill and in a TMP mill processing softwood, in order to establish i) the principal source of turpene losses in the kraft pulping process, and ii) the potential recovery rate of volatile terpenes in the TMP mill, as detailed below [25].

4.1. Investigation in a Kraft Pulp Mill

A French kraft pulp mill processing 1 125 000 T/y of maritime pine wood chips, from wood log chipping and sawmills, produces 350 000 ADT/y of UBSKP, and 4 kg/ADT of turpentine. The quantity of volatile terpenes in the initial wood chips, in the produced pulps, and in certain relevant gaseous samples was determined. The efficiency of the turpentine line was established on the basis of analyses of turpentine samples and process water samples (included condensates) (Figure 5).

Maritime pine wood chips sampled the day of the mapping study contained 0.2% of volatile terpenes, mainly α-pinene (0.14%) (Table 8). These contents were lower than those previously found on extra-fresh wood chips from logs (i.e., 0.6%) [13]. A lower terpene content would indicate that the materials studied would have a high ratio of sawmill residues [1], and would be derived from long-term stored wood logs. The two pulps produced contained very little volatile terpenes. The volatile terpenes in the kraft pulping process line were partly emitted in the vapor-output of the steaming/impregnation stage (ref. kT1, 230 kg/h of turpentine), and from the digester (ref. kT2, 310 kg/h of turpentine.).

During the steaming/impregnation stage, a significant amount of volatile terpenes was found in the effluent (16-22 kg/h). The black liquor evaporation plant, which produced sulfated turpentine, revealed significant turpene losses in the condenser effluents (9-11 kg/h). These uncaptured emissions represented up to 10% of the standard turpentine production of the mill, and contained especially large amounts of cymene, 3-carene and limonene, in addition to α-pinene and β-pinene compounds classically found. These process water side streams could explain the disruptions in the wastewater treatment plant and represent a significant economic loss.

4.2. Investigation in a TMP Mill

The French TMP mill investigated proceseses 250 000 T/y of wood and produces 700 ADT/d of TMP. The raw wood material is a mixture of spruce and fir wood chips supplied mainly by sawmills (75%) with the remainder made up of log chippings (25%). The wood chips are stored for around 3-4 weeks before entering the TMP process line. Wood chips are first washed at 75°C, and pre-steamed at 80°C, with steam at 80°C. During washing and steaming, waste vapor emissions are collected and evacuated through an extractor-blower (T_MG1, Figure 6). During the first and second refining stages, TMP waste steam (140°C, 3-4 bars) is scrubbed and condensed in a reboiler. The non-condensables are eliminated (T_MG3, Figure 6).

Softwood chips (mixture of spruce and fir) contained 0.05% of volatile terpenes, consisting of α-pinene (0.023%) and β-pinene (0.014%) (Table 9). These contents were lower than those previously found on extra-fresh wood chips from spruce and fir logs (i.e., 1.2% and 1.1% respectively) [13], meaning that the materials studied would contain a high proportion of sawmill residues [1]. The long storage time of wood chips before pulping might also explain these low terpene contents [1]. No significant volatile terpenes were found in the TMP. The volatile terpenes in the TMP process were mainly emitted in the vapor from the refining stage (ref. T_MG2 and T_MG3) with 2111 ± 131 mg/m^3 at the final outlet of the reboiler/scrubber device. The emissions from the steaming stage were lower, with a median flow of 3.5 kg/h (ref. T_MG1), as already indicated by Crawford and Jain [7] for total VOC emissions. The gas flow during sampling was not stable, which led to a wide variation in volatile terpene flow rate. A longer sampling duration time would no doubt give a more accurate evaluation. Almost no volatile terpenes were found in the process waters from the reboiler/scrubber.

The potential terpene emissions from wood in the TMP mill studied were estimated to be 15-30 kg/h based on initial contents of 0.05-0.1%. During the TMP process, the biggest source of volatile terpene was found in the air emissions from the refiner stream treatment, which was much higher than in the air emissions from steaming stage. In this mill, and under maximum production conditions, the turpentine production could reach 0.7 T/d. Moreover, to offer the possibility of additional income for the mill, the volatile terpenes could be trapped in order to reduce the VOC emissions of the TMP mill, and this would limit the amount of toxic components in the WWTP, thereby ensuring a positive environmental impact.

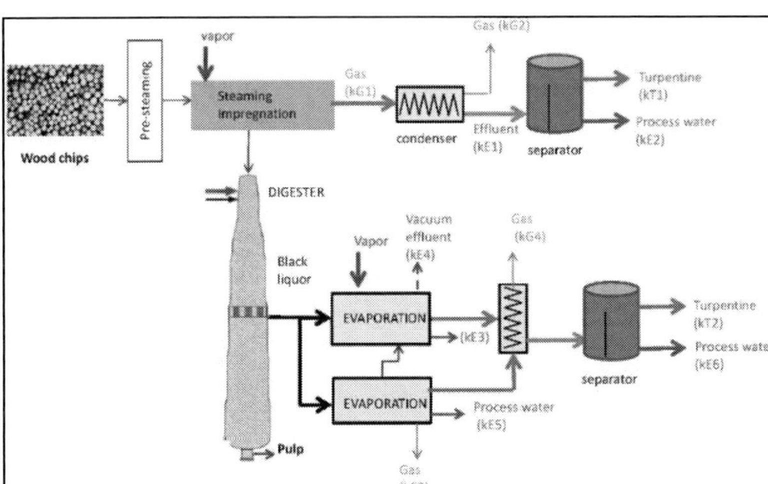

Figure 5. Block diagram of the process lines of the kraft pulp mill studied (sampled pulps: kP1, kP2, gaseous samples: kG1, kG2, kG3 and kG4, turpentine samples: kT1 and kT2 and process water samples: kE1, kE2, kE3, kE4, KE5, kE6) [25].

Table 8. Volatile terpene contents of wood chips, certain gaseous samples, process waters and turpentine in the turpentine line of the kraft pulp mill [25]

Terpenes	Wood chips	Gaseous stream		Turpentine		Process water		
	Maritime pine (45.7% dryness)	**kG1*** *6500-8000 m^3/h*	**kG2**** *52 m^3/h*	**kT1** *0.4 m^3/h*	**kT2** *0.5 m^3/h*	**kE2** *15-20 m^3/h*	**kE3** *12 m^3/h*	**kE5** *20 m^3/h*
	% (g/100g o.d.w)	**(mg/m^3)**	**(mg/m^3)**	**(g/m^3)**	**(g/m^3)**	**(g/m^3)**	**(g/m^3)**	**(g/m^3)**
α-pinene	0.143 ± 0.02	1 824 ± 212	201 ± 5	322 000	387 800	216	129	100
β-pinene	0.046 ± 0.005	546 ± 120	41 ± 7	157 500	155 500	62	55	29
Cymene	< 0.001	< 0.1	< 0.1	5 200	3 700	295	175	136
3-carene	< 0.001	< 0.1	< 0.1	< 1	< 1	239	142	110
Limonene	0.007 ± 0.0001	59 ± 25	< 0.1	27 200	25 000	268	159	124
Terpineol	< 0.001	< 0.1	< 0.1	31 600	26 200	<1	<1	<1
Caryophyllene	0.001 ± 0.0001	< 0.1	< 0.1	29 500	23 200	10	112	34
Total	**0.197 +/- 0.025**	**2 430 +/- 360**	**243 +/- 12**	**573 100** ***(230 kg/h)***	**621 300** ***(310 kg/h)***	**1 090** ***(16-22 k/h)***	**770** ***(9 kg/h)***	**534** ***(11 kg/h)***

*kG1: 98°C, 77% relative humidity; ** kG2: 29.6°C, 94.3% relative humidity

Table 9. Volatile terpene contents of wood chips and gaseous samples along the process line of the TMP mill [25]

Terpenes	Wood chips	Gaseous stream			Process water	
	Spruce + fir (46% dryness)	T_MG1* *3 500 – 17 700 m³/h*	T_MG2** *15 000 – 34 000 m³/h*	T_MG3*** *3 000 – 15 400 m3/h*	T_ME1 *3.6 m3/h*	T_ME2 *19.8 m³/h*
	% (g/100g o.d.w)	mg/m3	mg/m3	mg/m3	mg/m3	mg/m3
α-pinene	0.023 ± 0.001	193 ± 11	2 196 ± 253	1 115 ± 64	28	1
β-pinene	0.014 ± 0.0001	110 ± 0.6	2 470 ± 909	719 ± 54	28	<0.5
Cymene	< 0.0001	< 0.1	< 0.1	< 0.1	< 0.5	<0.5
3-carene	0.003 ± 0.0001	7 ± 1.6	700 ± 310	100 ± 111	3	<0.5
Limonene	0.009 ± 0.001	11 ± 1.1	1 447 ± 1 300	163 ± 36	5	<0.5
Terpineol	< 0.0001	2 ± 1.1	267 ± 236	4 ± 3	<0.5	<0.5
Caryophyllene	< 0.0001	2 ± 0.8	187 ± 250	10 ± 3	1	<0.5
Total	**0.049 +/- 0.002**	**325 +/- 13** ***(1 -6 kg/h)***	**7 267 +/- 2 282** ***(75 -330 kg/h)***	**2 111 +/- 131** **6 – 35 kg/h**	**64** ***(0.0002)***	**1 (*<0.00001*)**

*T_MG1: 62°C, 79% relative humidity; ** T_MG1: 99.8°C, 96% relative humidity; ***T_MG3: 98°C, 38% relative humidity

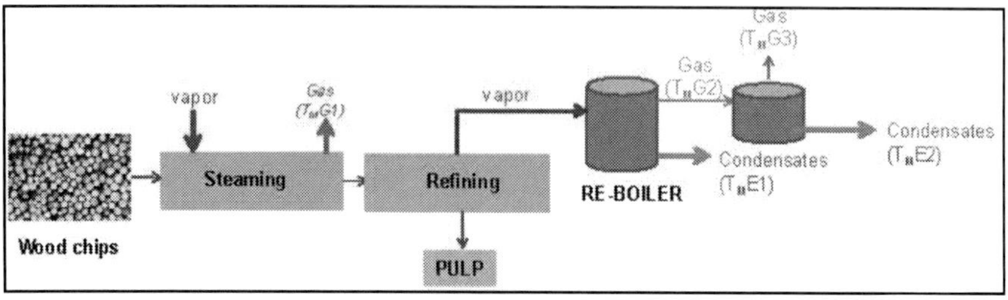

Figure 6. Block diagram of the TMP mill process: identification of the samples studied (gaseous samples: T_MG1, T_MG2, T_MG3 and process water samples: T_ME1, T_ME2) [25].

ACKNOWLEDGMENTS

Financial support for part of this research was provided by CTPi members and the French Ministry of Industry. The authors would also like to thank the two French pulp mills for their financial and technical support in this study.

REFERENCES

[1] Strömvall, A. M. & Petersson, G. (2000). *Volatile terpenes emitted to air. Pitch control, wood resin and deresination*, edited by: Back, EL and Allen, LH, TAPPI Press, Atlanta, 77-99.

[2] Sixta, H., Potthast, A. & Krotschek, W. (2006). Chemical pulping processes, In Sixta, Herbert, Ed. *Handbook of pulp.*, Vol. *1*. Weinheim: Wiley-vch, 109-510.

[3] Gullichsen, J. & Lindberg, L. (1999). By-products of chemical pulping, Chapter 18. In: Gullichsen J. & Paulapuro H. Eds. *Chemical pulping Book 6B*, Helsinki: Tappi Press, 1999, 374-389.

[4] Blanchy, M. & Alcorta, J. (2013). *La filière de la résine de pin en, 2013, projet SUSt FOREST*, www.sust.forest.eu. *[The pine resin industry in 2013, SUSt FOREST project*, www.sust.forest.eu]

[5] Silvestre, A. J. & Gandini, A. (2008). Terpenes: major sources, properties and applications. *In Monomers, Polymers and Composites from Renewable Resources*, (pp. 17-38).

[6] Capital, T. V. (2004). Part 3. *Aroma Chemicals Derived from Effluent from the Paper and Pulp Industry*, FRIDGE, NEDLAC, South Africa.

[7] Crawford, R. J. & Jain, A. K. (1997). Volatile organic compound emissions from non-chemical pulp and paper mill sources – Part III mechanical pulping. In: *NCASI technical Bulletin N°738*, July 1997.

[8] Rauscher, J. W. (2002). Modern TMP heat and turpentine recovery. *In Annual Meeting – Pulp And Paper Technical Association of Canada* (Vol. *88*, No. A, pp. A91-A94). Pulp and Paper Technical Association of Canada, 1999.

[9] Fengel, D. & Wegener, G. (1984). Wood: chemistry, ultrastructure, reactions. *Walter de Gruyter*, *613*, 1960-82.

[10] Back, E. L. (2000). *The locations and morphology of resin components in the wood. Pitch control, wood resin and deresination*, 1st ed., TAPPI, Atlanta, GA, 1-28.

[11] Alén, R. (2000). Structure and chemical composition of wood. In: Gullichsen J. & Paulapuro H Eds. OyHelsinki: Fapet Oy, 2000. *Forest products chemistry*, Chapter 1, 11-57.

[12] Ekmann, R. & Holmbom, B. (2000). The chemistry of wood resin, In Allen E. B. Ed. *Pitch control, wood resin and deresination*. Atlanta, USA: Tappi Press, 2000, 37-76.

[13] Bertaud, F., Crampon, C. & Badens, E. (2017). Volatile terpene extraction of spruce, fir and maritime pine wood: supercritical CO_2 extraction compared to classical solvent extractions and steam distillation. *Holzforschung*, *71*(7-8), 667-673.

[14] Strömvall, A. M. (1992). *Terpenes emitted to air from forestry and the forest industry*. Doctoral Thesis, Chalmers University of Technology, Sweden.

[15] Granström, K. (2005). *Emissions of volatile organic compounds from wood* Doctoral dissertation., Karlstad Universitiy, Sweden.

[16] RISI. (2013). *Statistics world pulp*, 2011.

[17] Alèn, R. (2000). Basic chemistry of wood delignification, In *Forest products chemistry*, Gullichsen J. and Pauluporuro H. Eds, 58-104.

[18] Gullichsen, J. & Lindeberg, H. (1999). By-products of chemical pulping, In Gullichsen, Johan, and Carl-Johan Fogelholm Eds. *Chemical pulping*., (Vol. *2*) Fapet Oy, B375-B389.

[19] Holmbom, B. (2011). Extraction and utilisation of non-structural wood and bark components. In *Biorefining of forest resources*, 178-224.

[20] CEPI. (2014). *Pulp and paper Industry – Definition and Concepts*, Cepi.

[21] Franzén, R. (1986). General and selective upgrading of mechanical pulps, *Nordic Pulp Paper Res. J.*, *3*, 4-13.

[22] Karnis, A. (1994). The mechanism of fiber development in mechanical pulping, *J. Pulp Paper Sci.*, 20(10), J280-J288.

[23] Sueiro, L. & Gill, J. (1995). Control of VOC emissions from a TMP plant. In *The 1995 International Environmental Conference*. Part 2(of 2), Atlanta, GA, USA, 05/07-10/95, (pp. 925-930).

[24] Crawford, R. J. & Jain, A. K. (1997), Volatile organic compound emissions from non-chemical pulp and paper mill sources, part III-Mechanical pulping, *Technical Bulletin* (No738). NCASI.

[25] Bertaud, F., Chareyre, P. & Petit-Conil, M. (2017). Volatile terpenes emission in pulp mills: investigation of a kraft pulp mill and potential recovering in a TMP mill, *Proceedings of the Nordic Wood BioRefinig Conference*, *Stockholm*, March 28-30 2017.

[26] Suckling, I. (2012). *Mechanical pulping and pulp bleaching chemistry*, *APPI*, May 2012 https://www.slideserve.com/kirby/mechanical-pulping-and-pulp-bleaching-chemistry.

[27] Sarwar, A. (2012). *Plant design for the separation of various components from turpentine oil*. Master of Science Thesis, Chalmers University of Technology, Göteborg, Sweden, February 2012.

[28] Rockhill, R. G. & Hostetter, D. W. (1982). Evaporator/reboiler recovers steam, turpentine from TMP refiners, *Pulps & Paper*, October 1982, 106-108.

[29] Rauscher, J. W. (2003). Modern TMP heat and turpentine recovery: Today, system operation is virtually trouble-free. *Pulp & paper* Canada, *104*(4), 26-28.

[30] Ryham, R. (1990). U.S. Patent No. 4,925,527. Washington, DC: U.S. Patent and Trademark Office.

[31] Weston, R. J. (2007). Recovery of turpentine from the production of medium-density fibreboard. *New Zealand Journal of Forestry Science*, *37*(1), 124.

[32] Coppen, J. J. (1995). *Non-wood forest products 1: flavors and fragrances of plant origin.* Rome: Food and Agriculture Organization of the United Nations x, 101p. ISBN, 661101897.

[33] Ukkonen, K. A. (2016). *Pine chemicals – Global View*, Pine Chemical Association International Conferences, Santiago, Chile, September 19, 2016.

[34] Borg-Karlson, A. K., Lindström, M., Norin, T., Persson, M. & Valterová, I. (1993). Enantiomeric Composition of Monoterpene Hydrocarbons in Different Tissues of Norway Spruce, Picea ab/es (L) Karst. A Multi-dimensional Gas Chromatography Study. *Acta Chemica Scandinavia*, *47*, 138-144.

[35] Gullichsen, J. & Lindeberg, H. (1999). By-products of chemical pulping, In Gullichsen, Johan, and Carl-Johan Fogelholm Eds. *Chemical pulping.*, (Vol. *2*). Fapet Oy, B375-B389.

[36] Leseche, B., Levisalles, J. & Rudler, H. (1984). L'essence de térébenthine du pin maritime: *Pinus pinaster. Parfums, cosmétiques, arômes*, (*58*), 53-58 [Gum turpentine of Maritime pine: *Pinus pinaster. Parfums, cosmétiques, arômes*, (*58*), 53-58].

[37] Jain, A. K., Crawford, R. J., Dillard, D. S. & et Rovell-Rixx, D. S. (1994). Volatile organic emissions from pulp and paper mill sources, Part IV, *Technical bulletin* N°0678.

Biographical Sketch

F. Bertaud/Frédérique Bertaud

Affiliation: Paper Technical Center, Grenoble, France

Education: PhD Thesis

Business Address: Centre Technique du Papier (CTP), InTechFibres Dpt. CS 90251, F-38044 Grenoble cedex 9, France

Research and Professional Experience:

Research Engineer at the CTP since 2001, she is a Project Manager in the 'InTechFibres-Plant Chemistry' team led by Mr Michel PETIT-CONIL. She is in charge of the "Bioraffinery" activities associated with pulp mills. Upon strong expertise in chemical analysis, she has become an expert in the extraction of wood-extractive compounds (polyphenols, tannins, terpenes) with the development of numerous analytical tools, process optimization, and the study of The value chain for the valorization of wood by-products. Experimented in the chemical pulping processes, she has also developed in CTP, the valorization of lignins from industrial black liquors, in partnership with her colleagues of the FCBA, and in close collaboration with industry and research laboratories. Having experience as a project coordinator (European and national projects), she is regularly responsible for the building and follow-up of projects, as well as the valorization of results through publications, communications and development of new lines of expertise of CTP.

Honors: Coordinator of the EU project 'CERBERUS' QLK QLK5-CT-2002-01027, qualified as 'Success story' by the EU

Publications from the Last 3 Years:

Journals

Bertaud F., Guillemain A., Tapin-Lingua S., Molina-Boisseau S., Petit-Conil M. (2015) Softwood Kraft lignin isolation: from laboratory to pilot-scale production, *Revue Atip*, 69 (3), 6-9.

Bertaud, F., Crampon, C., Badens, E. (2017) Volatile terpene extraction of spruce, fir and maritime pine wood: supercritical CO2 extraction compared to classical solvent extractions and steam distillation. *Holzforschung*, *71*(7-8), 667-673.

Serrano Cantador, Luis; Esakkimuthu, Esakkiammal Sudha; Marlin, Nathalie; Brochier-Salon, Marie-Christine; Mortha, Gerard; Bertaud, Frederique (2018) Fast, easy and economical quantification of lignin phenolic hydroxyl groups. Comparison with classical techniques", *ACS Sustainable Chemistry & Engineering Manuscript ID: sc-2017-04479q*

Conferences

Bertaud F. (2015). Valorisation des composés polyphénoliques et autres substances extractibles du bois: une valeur ajoutée aux filières bois, pâtes et papiers, *Colloque Atelier 7: 3ème Forum Franco-Québécois InTechFibres*, in Québec Canada, 2-4 nov. 2015.

Bertaud F., Crampon E., Badens E., Petit-Conil. M. (2016) Volatile terpene extraction of spruce, fir and maritime pine wood: supercritical CO2 compared to classical solvent extractions and steam distillation EWLP - *European Workshop on Lignocellulosics and Pulp in Autrans*, France, June 28-July 1 2016.

Bertaud F., Lichère H., Serrano L, Mortha G., Marlin N. (2017) Determination of functional groups of industrial lignins as a key indicators for their use as novel green ingredients, *ISGC - 4th International Symposium on Green Chemistry, Flash Presentation* in La Rochelle, France, 16-19 may 2017.

Bertaud F.; Dufour A.; Mauviel G. (2017) Production de composés phénoliques par liquéfaction de la lignine, *Journée de valorisation du projet Fui Polywood,* Axelera in France, Grenoble 10 mai 2017, CTP - Centre Technique du Papier.

Bertaud F.; Petit-Conil M. (2017) Terpènes, colophane et TOFA extraits du bois : une valeur ajoutée pour les usines de pâtes à papier, *Journée de valorisation du projet Fui Polywood*, Axelera in France, Grenoble 10 mai 2017, CTP -Centre Technique du Papier.

Serrano L.; Esakkimuthu S.; Marlin N.; Bertaud F.; Mortha G. (2017) Fast method for lignin phenolic hydroxyl groups quantification. Comparison with classical techniques, *10th World Congress of Chemical Engineering in Espagne, Barcelone* 01-05 Oct. 2017.

Serrano L.; Esakkimuthu S.; Marlin N.; Bertaud F.; Mortha G. (2017) Fast vs classical methods for phenolic hydroxyl groups quantification., *4th Workshop MC Meeting Cost FP1306 valorization of lignocellulosic biomass to Chemicals, Materials and Fuels in Malte, Torremolinos* 27-29 March 2017, COST - European Cooperation in the Field of Scientific and Technical Research.

Bertaud F, Petit-Conil M., Lichère H., Mortha G. (2017) Determination of sulfonic and carboxyl acids contents of industrial lignosulfonates, Poster, *7th Nordic WoodBiorefinery Conference*, in Stockholm, Sweden, 28-30 March, 2017, 2 p.

In: Terpenes
Editor: Albin Bjarke
ISBN: 978-1-53613-622-7

Chapter 4

EFFECTS ON THE CARDIOVASCULAR SYSTEM OF TERPENE-RICH ESSENTIAL OILS

Jorge Alves-Silva[1,2], Júlia Barbeitos[1,2], Carla Marques[2], Lígia Salgueiro[1], Henrique Girão[2] and Mónica Zuzarte[2,*]
[1]CIEPQPF and Faculty of Pharmacy, University of Coimbra, Coimbra, Portugal
[2]iCBR and Faculty of Medicine, University of Coimbra, Coimbra, Portugal

ABSTRACT

Cardiovascular diseases remain a global burden with tendency to increase due to behavioural risk factors including tobacco and alcohol abuse, unhealthy diet, obesity, and physical inactivity. Current treatments include clinical procedures and medications that are clearly not sufficient, thus justifying novel strategies that may include preventive approaches, such as a healthy lifestyle. Indeed, it is widely accepted that a Mediterranean-style diet, rich in plant intake, can have cardioprotective

[*] Corresponding Author Email: mzuzarte@uc.pt.

effects by attenuating cardiac risk factors and/or modulating disease's onset and progression. In fact, several compounds present in plants, namely secondary metabolites, such as terpenes, have broad therapeutic effects with several studies ascribing beneficial cardiac effects. Bearing in mind that plant essential oils are complex mixtures of several compounds, mainly terpenes of low molecular weight (monoterpenes and sesquiterpenes), the present review focuses on the effects of terpene-rich essential oils. Although several studies describe the cardioprotective effects of these essential oils, a systematic review is lacking. Therefore, a compilation of these studies is presented with special focus on the hypotensive, vasorelaxant, modulation of ion channels, anti-dyslipidaemic, and antiplatelet aggregation effects as well as the protective effects against myocardial infarction. Moreover, putative mechanisms of action involved in these effects are presented as well as an attempt to establish a composition-activity relationship.

Keywords: volatile extracts, cardioprotective, hypotensive, vasorelaxant, ion channel modulation, anti-dyslipidaemic, antiplatelet aggregation

1. Introduction

According to the World Health Organization (WHO), cardiovascular diseases (CVDs) account for 31% of deaths worldwide [1]. Aside aging, these diseases are triggered by a plethora of factors, that are very common in the modern world such as, tobacco, alcohol abuse and unhealthy diets. These risk behaviours lead to high blood pressure, overweight and several other complications that promote the onset of early CVDs. Indeed, the WHO has launched three technical packages aiming to decrease the risk of these disorders, namely MPOWER for tobacco control, SHAKE to reduce salt consumption and HEART for strengthening of CVDs' management in primary health care [2]. Moreover, therapeutic strategies and surgical interventions are used to improve patients' quality of life and contribute to their long-term survival. Nevertheless, bearing in mind the burden of CVDs, other approaches have gain interest among the scientific and medical community, namely plant-based strategies [3]. Despite the development of chemically synthesized drugs, approximately 80% of the world's population still relies on herbal medicines as a source of primary health care [4]. Also, the Mediterranean plant-based diet has been greatly adopted due to its known health benefits directly on CVDs or indirectly by reducing associated risk factors such as atherosclerosis and high cholesterol [5]. In fact, the Lifestyle Heart Trial compared a plant-based diet with a regular diet and showed that 82% of the enrolled patients with heart disease had some level of regression of atherosclerotic plaque and 91% had a reduction of angina episodes when subjected to a plant-based diet, whereas 53% of the control group had a progression of the atherosclerotic plaque [6]. Currently, herbs are also used in the treatment of chronic and acute conditions such as CVDs, prostate problems, depression, inflammation, and to boost the immune system [7]. These effects are generally attributed to the presence of secondary metabolites, such as terpenes [8]. Importantly, these compounds are relevant in drug development, being used directly as therapeutic agents, or as starting materials and models for the synthesis of other drugs [7].

Previously, a review on the effects of terpenes on the cardiovascular system highlighted a wide array of protective effects directly on this system or indirectly through the modulation of major risk factors associated with these diseases [9]. Therefore, it is conceivable that plant extracts, namely essential oils rich in terpenes, may have beneficial cardiac effects that can be higher than that of the isolated compounds. For this reason, in this chapter, a compilation of the cardioprotective effects of essential oils rich in terpenes is presented. The different classes of CVDs are summarized with focus on their major risk factors and current treatments, followed by a brief characterization of essential oils. Finally, an extensive review on the cardioprotective effects of essential oils is presented with focus on their hypotensive, vasorelaxant, ion channel modulation, antiplatelet aggregation and anti-dyslipidaemic potential. Also, the effects of the isolated terpenes is pointed out and, whenever known, the mechanism of action associated to the protective effect is referred and a composition-activity relationship established.

The research was conducted on different search engines, namely *Pubmed*, *Google Scholar* and *Scopus* using combinations of keywords such as « essential oil » and « cardiovascular system ». Papers with essential oils with high content of terpenes that were published between 2000 and 2018 were considered.

2. Cardiovascular Diseases

2.1. General Considerations

Cardiovascular diseases are one of the leading causes of mortality worldwide [10], accounting for more than 4 million deaths each year in Europe alone [11]. Due to the rising age of the population, the incidence of CVDs tends to increase and, for this reason, the need for novel and efficacious therapeutic strategies is imperative [12]. In 2015, it was estimated that 17.7 million people died from CVDs, being 7.4 million due to coronary artery disease and 6.7 million due to stroke [1]. Moreover, these disorders also affect the quality of life of patients, thus accounting for 10% of the global disease burden [12,13].

The WHO describes CVDs as a group of disorders that affect the heart and blood vessels, such as coronary artery disease (angina and heart attack), cerebrovascular disease (where normal blood supply to part of the brain is cut off), peripheral arterial disease (narrowing of arteries usually in the legs), rheumatic heart diseases, congenital heart diseases and deep vein thrombosis [14]. Atherosclerosis is the major cause of CVDs and refers to the narrowing of the arteries due to plaque build-ups (cholesterol deposits) in the inner wall [15]. Moreover, the most common causes of cardiovascular death are coronary artery disease and cerebrovascular disease [11].

Table 1 summarizes the 6 classes of CVDs according to the WHO [1] and points out the body region mostly affected, associated pathophysiology as well as the prevalence/incidence and consequent mortality, according to specific guidelines.

Table 1. Cardiovascular Diseases

Cardiovascular Disease	Body region	Pathophysiology	Prevalence/Incidence	Mortality	Ref
Coronary Artery Disease	Coronary artery	Disease of the blood vessels supplying the heart muscle	Prevalence of angina: 5-7% (♀), 4-7% (♂), 45-64 years old 0-12% (♀), 12-14% (♂), 65-84 years old	1.2 – 2.4% *per* year	[16]
Peripheral Artery Disease	Carotid, mesenteric, renal, upper and lower extremities arteries	Disease of blood vessels supplying peripheral arteries	Carotid artery: 2.2% (♀), 4.8% (♂), <70 years old 6.9% (♀), 12.5% (♂), >70 years old	63% of patients with >50% asymptomatic carotid stenosis showed late death associated with cardiac events	[17]
			Upper extremities artery: 2%, 9% with concomitant LEAD		
			Mesenteric artery: 14% in patients with routine cardiac catheterization		
			Renal artery: 4.1% patients without CAD 9.9% patients with CAD		
			Lower extremities arteries: 202 million patients worldwide, 40 million patients in Europe		
Cerebrovascular Disease	Brain	Disease of the blood vessels supplying the brain	Incidence: 66 – 220/100000/year	14% (♀), 9% (♂)	[10,18]
Rheumatic Heart Disease	Muscles and valves of heart	Damage to the heart muscle and heart valves from rheumatic fever, caused by streptococcal bacteria	Prevalence: 444/100000 in countries with an endemic pattern 3.4/100000 in countries with no endemic pattern	4.8 deaths/100000 habitants	[19]
Congenital Heart Disease	Heart	Malformations of heart structure existing at birth	~2800/1000000 habitants	77% of patients with ACHD	[20,21]
Deep vein thrombosis and pulmonary embolism	Veins in the lower limbs	Venous thrombus which can dislodge and move to the heart and lungs	70 - 140/100000 person-year	Short-term mortality 2 – 5%.	[22]

♀ - women; ♂ - men; ACHD – adult congenital heart disease; CAD – coronary artery disease; LEAD – lower extremities arterial disease

2.2. Risk Factors

Several risk factors are responsible for the onset and development of CVDs, such as aging, high blood pressure, high cholesterol levels, diabetes, obesity, tobacco, alcohol, and physical inactivity. Risk factors associated with age, sex or genetic predisposition are considered nonmodifiable whereas those that depend on lifestyle, are generally referred to as modifiable risk factors [1,23].

Within the first group, family history of heart-related problems plays a crucial role in the CVDs' onset. Moreover, genetic associations with type 2 diabetes, hypertension, high cholesterol and abnormal blood lipids, also increase the risk of these disorders [23]. Other common nonmodifiable factors include socioeconomic status, age, gender and ethnicity [23]. For example, men before the age of 55 have a higher risk for CVDs than women, but following menopause, women are more prone to develop these disorders [24]. Also, African and Asian ethnicities have a higher risk of developing CVDs [25].

Regarding modifiable risk factors, diet and physical activity play crucial roles. Indeed, high saturated fat-diets increase the risk of heart disease and stroke [26]. It is known that saturated and trans fats lead to high levels of cholesterol whereas, unsaturated fats are associated with a protective and beneficial effect on the cardiovascular health [27]. Increased blood cholesterol is a decisive risk factor for CVDs, and in the presence of other factors, such as high blood pressure and tobacco smoke, the risk is increased. In order to reduce high cholesterol, as well as elevated blood pressure and type 2 diabetes, a healthy diet is recommended [28–30]. Furthermore, high salt consumption and low fruit and vegetable intake also increase the risk of acute myocardial infarction [31].

On the other hand, physical activity is equally beneficial since it reduces CVDs associated mortality [32]. The WHO believes that more than 60% of the global population is not sufficiently active. This is very relevant since the protective effect of physical activity includes weight regulation and insulin sensitivity improvement, maintenance of blood pressure, blood lipid and glucose levels and ameliorates the health of blood vessels [23,33].

Another major risk factor is smoking that increases heart rate and blood pressure, thus promoting abnormal heart rhythms, alterations in cholesterol levels and damages to blood vessels [34]. Moreover, excessive alcohol can lead to an increase in blood pressure, with concomitant arrhythmias, as well as an increase in high levels of triglycerides, thus leading to obesity [35].

In the last decade, an increased emphasis on patient education aiming the reduction of CVDs risk factors has been accomplished, by promoting a healthy lifestyle, that includes avoiding smoking, eating a low fat diet, exercising regularly and maintaining a healthy weight [23].

2.3. Management and Treatment

The European Society for Cardiology (ESC) designed specific guidelines for the prevention of CVDs that address modifiable risk factors, namely changes in lifestyles [36]. Accordingly, two recent clinical trials demonstrated that an healthy diet (Mediterranean-style diet) in combination with other lifestyle factors, like physical activity, moderate alcohol consumption and no smoking, decreased the all-cause mortality in 10 years and improved quality of life [37,38]. Nevertheless, when CVDs are already installed, pharmacological and/or surgical approaches are needed.

For coronary artery disease (CAD), the available drug classes comprise anti-ischaemic, antiplatelet, and lipid-lowering agents as well as renin-angiotensin-aldosterone system blockers and analgesic drugs. The anti-ischaemic drugs are pleiotropic and include for example short- and long-acting nitrates that induce coronary vasodilation and β-blockers which have a direct effect on the heart by decreasing heart rate, contractility, atrioventricular conduction and ectopic activity. Furthermore, the protective effect of these drugs may also be due to an increase in the perfusion of the ischaemic area. Antiplatelet agents such as low-doses of aspirin or P2Y12 inhibitors as well as statins, widely used for the management of dyslipidaemia, are also employed to control cardiovascular events in patients. Angiotensin converting enzyme (ACE) inhibitors are known to decrease myocardial infarction, stroke and heart failure and, therefore, are also applied in CAD events. Calcium channel blockers achieve their pharmacological effect by inhibiting L-channel opening in the vascular smooth cells and myocardium, thus causing vasodilation and reduction of the peripheral vascular resistance. In addition to the described treatments, CAD can also be improved by revascularization of the coronary arteries by percutaneous coronary intervention or by coronary artery bypass surgery [16].

For the management of peripheral artery disease, hypotensive drugs are advised, although caution is needed to avoid a decrease in systolic blood pressure to values below 110 – 120 mmHg. For patients with carotid artery disease, clopidogrel or aspirin, used as antithrombotic agents, can be used. Regarding patients with lower extremities arterial disease (LEAD), antiplatelet therapies can also be employed, however specific indications are still unknown since no study has been conducted to assess the effect of these therapies on the full spectrum of LEAD. Recently, the Cardiovascular OutcoMes for People Using Anticoagulation StrategieS (COMPASS) trial assessed the efficacy of rivaroxaban or aspirin monotherapies with the dual antiplatelet therapy (rivaroxaban + aspirin). The main outcomes highlighted the benefits of the use of rivaroxaban + aspirin in decreasing limb- and cardiovascular-associated events [39]. In addition to the pharmacological approaches, peripheral artery disease (PAD) can also be managed by revascularization or endovascular surgery [17].

According to the ESC, cerebrovascular disease is a consequence of two types of PAD, namely carotid artery disease and vertebral artery disease [17]. The first accounts for most of the cerebrovascular events, while the second is responsible for about 20% of the ischaemic cerebrovascular events associated with posterior circulation. Therefore, the management of this disease is the same as described previously for PAD.

According to the ESC/European Association for Cardio-Thoracic Surgery (EACTS) guidelines, valvular heart disease (VHD) include aortic, mitral or tricuspid regurgitation, aortic, mitral or tricuspid stenosis and combined and multiple valve disease [19]. Despite this heterogeneity, the management of VHD consists mainly of surgical interventions, namely valve repair or replacement. However, in some cases pharmacological approaches can also be employed, being angiotensin-converting-enzyme inhibitors and β-blockers, the main classes used.

Due to cardiac defects' heterogeneity in congenital heart disease, the management of this disorder is dependent on the type of defect [20]. For example, an atrial septal defect can be repaired by surgery, with a better outcome in younger patients when compared to elder ones. Nevertheless, all patients benefit from septal repair in terms of morbidity although, in some cases, complementary treatments are required. Similarly, ventricular septal defect is also mainly corrected by surgery via pericardial patch. However, in patients with high surgical risk, transcatheter closure is advisable.

The management of deep vein thrombosis comprises three phases [22]. In the first two phases, anticoagulation therapy is applied, being the main classes of drugs used vitamin K antagonists or direct oral anticoagulants. Furthermore, thrombolysis or thrombectomy can also be employed. When anticoagulation therapy is not feasible, vena cava filters can be applied however, several disadvantages have been described for this method. When high risk of pulmonary embolism exists, anticoagulation therapy followed by reperfusion is the preferred treatment, while in patients with moderate risk, anticoagulation therapy alone is applied. Nevertheless, if haemodynamic decompensation occurs, reperfusion can be used as a rescue treatment. In patients with isolated distal deep vein thrombosis, a full-dose anticoagulation is done for 3 months with follow-up via ultrasound, to monitor thrombus progression. Furthermore, the use of compression stockings is recommended in all patients.

Importantly, these therapies are associated with several adverse effects that are dependent of the individual, the disease being treated and the exposure to other drugs due to drug-drug interactions [40].

3. Essential Oils

3.1. General Aspects

The European Pharmacopeia defines essential oil as an "odorous product, usually of complex composition, obtained from a botanically defined plant raw material, by steam distillation, dry distillation, or a suitable mechanical method without heating" [41]. Essential oils can be found in several organs of aromatic plants being produced and stored in specialized secretory structures that can be found on the surface of the plant organs or within its tissues [42]. Most of the methods applied in the analysis of essential oils rely on chromatographic procedures, which enable component separation and identification, being the most used methods gas chromatography (GC) and gas chromatography/mass spectrometry (GC/MS) [43,44]. In general, the identification of the compounds present in essential oils is made by the comparison of both the GC retention indices and mass spectra data with those of authentic samples [45,46]. For more complex mixtures, several fractionating techniques are applied and to identify compounds that are not easily separated by GC or essential oils with similar stereo-isomeric compounds, other methodologies can be used such as gas chromatography/Fourier Transform Infrared (GC-FTIR) and Nuclear Magnetic Resonance Spectroscopy (^{13}C-NMR) [47]. Essential oils' composition may vary due to both intrinsic (sexual, seasonal, ontogenetic and genetic variations) and extrinsic (ecological and environmental aspects) factors [48,49]. Importantly, genetic variations may result in the expression of different metabolic pathways and consequently, quantitative and qualitative variations in essential oil's composition may occur. When significant differences are found, an intraspecific category (chemotype) is identified [48]. Analytical guidelines published by several institutions such as European Pharmacopoeia, International Standards Organization and WHO are available and must be followed to assure the good quality of the commercialized essential oils and of the plants from which they are obtained.

3.2. Chemical Composition and Biological Effects

Terpenes are one of the main groups of compounds found in essential oils. However, other compounds namely, phenylpropanoids, fatty acids and their esters and, more rarely, nitrogen and sulphur derivatives may occur [50–52]. In aromatic plants, terpenes are biosynthesized *via* two pathways: the mevalonate dependent (MVA) pathway that takes place in the cytosol and the mevalonate independent or 2C-metil-D-erythritol 4-phosphate/1-deoxi-D-xilulose-5-phosphate (MEP/DOXP) pathway that occurs in the chloroplasts [53]. Terpenes result from the condensation of a pentacarbonate unit (isoprene) being classified according to the number of isoprene units in their structure, as hemiterpenes (1 unit), monoterpenes (2 units), sesquiterpenes (3 units), diterpenes (4 units) and so on. Terpenes of low molecular weight, like monoterpenes ($C_{10}H_{16}$) and sesquiterpenes ($C_{15}H_{24}$) are the most predominant in essential oils. These compounds have many isomeric cyclic or linear structures, various degrees of unsaturations, substitutions, and oxygenated derivatives.

Aromatic plants and their products have long been used by local communities mainly as bactericidal, virucidal, fungicidal, anti-inflammatory, anti-parasitical and insecticidal agents [50]. However, many uses are based on old beliefs or faiths usually without any scientific support of their efficacy. This situation may cause severe problems and must be avoided through the identification of plant properties on a scientific-based manner. Indeed, in recent years plant secondary metabolites, including essential oils have attracted the attention of many investigators, with several studies corroborating their traditional uses or identifying new potential therapeutic targets. Moreover, essential oils are valuable natural products used as raw materials in many fields, like pharmaceutical, agronomic, food, sanitary, cosmetic and perfume industries [54].

Importantly, many plant families such as Apiaceae, Asteraceae, Cupressaceae, Lamiaceae, Lauraceae, Myrtaceae, Pinaceae and Rutaceae comprise well-known aromatic species, many of them included on the Generally Recognized as Safe (GRAS) list fully approved by the US Food and Drug Administration and Environmental Protection Agency for addition to food and beverages. Moreover, the Committee on Herbal Medicinal Products (HMPC) has disclosed several monographs on aromatic plants and their essential oils (traditional use and well-established use) [55] and some of these oils can be found in commercially available drugs.

4. Effect of Terpenes-Rich Essential Oils and Isolated Terpenes on the Cardiovascular System

During the last years, there has been a revived interest in the use of aromatic and medicinal plants. Interestingly, these plants and their extracts have been used for centuries to treat cardiovascular-related disorders such as congestive heart failure, systolic hypertension, angina pectoris, atherosclerosis, cerebral insufficiency, venous insufficiency and arrhythmia [56]. Moreover, beneficial effects in the attenuation of major risk factors, such as diabetes, hypertension and high levels of cholesterol, have also been reported [3]. Importantly, many of these effects are due the presence of secondary metabolites, such as terpenes. Indeed, a recent review carried out by our group compiled the protective effects reported for terpenes on the cardiovascular system [9]. Now we intend to gather the information available on essential oils with high amounts of these compounds. Therefore, in the following tables, a systematic review on the cardioprotective effects of these extracts is presented with special focus on their hypotensive, vasorelaxant, modulation of ion channels, anti-dyslipidaemic, and antiplatelet aggregation effects as well as protection against myocardial infarction. In addition, the effect of isolated terpenes frequently found in essential oils is also described and, whenever known, the putative mechanism of action underlying these properties as well as a composition – activity relation is pointed out.

4.1. Hypotensive Effects

Hypertension, or high blood pressure, is a long-term medical condition characterized by a systolic blood pressure (SBP) ≥ 140 mmHg and/or diastolic blood pressure (DBP) ≥ 90 mmHg [57]. Depending on the real values of SBP and DBP, hypertension can be classified into 3 grades: Grade I – SBP: 140-159 and/or DBP: 90-99; Grade II – SBP: 160-179 and/or DBP: 100-109 and Grade III – SBP: ≥180 and/or DBP: ≥110. This disease affects 30 – 45% of the population with a steep increase with age. Due to the prevalence of hypertension and the concomitant presence of several risk factors which lead to an increased mortality risk in 10-years' time (SCORE project) [58], an efficient management is imperative. Antihypertensive drugs namely diuretics, β-blockers, calcium antagonists, ACE inhibitors, and Angiotensin II Receptor Blockers (ARBs) are able to decrease blood pressure. However, about 35% of hypertensive patients discontinue their medication within 6 months, and about 50% of cases show adverse effects, such as the risk of hyperkalaemia with diuretics or oedema when using calcium antagonists [59], thus justifying the search for effective alternative approaches. Regarding plant extracts, several *in vivo* studies have described the hypotensive effects of terpene-rich essential oils as summarized in Table 2.

Despite *in vivo* evidence on the hypotensive efficacy of several essential oils, their mechanisms of action are not very clear. Nevertheless, for *Mentha* x *villosa* essential oil, it was shown that the hypotensive effect occurs *via* a NO-dependent pathway and the bradycardic effect was independent from the autonomic nervous system [67].

The effect of most of the reported essential oils is due to the presence of active terpenes, namely hydrocarbon or oxygenated monoterpenes. Regarding hydrocarbon monoterpenes, (+)-α-pinene and (-)-β-pinene, were reported as inductors of hypotension associated with tachycardia in a dose-dependent manner [71]. For oxygenated monoterpenes, several studies have been carried out using mainly rat models. For example, linalool induced hypotension associated with tachycardia [71,72] and decreased the mean arterial pressure without affecting heart rate.

Table 2. Essential oils with hypotensive effects

Essential Oil	Part of Plant	Terpenes	Study Model	Cardiovascular Effect	Ref.
Alpinia zerumbet	Leaves	Terpinen-4-ol (28.1%), 1,8-Cineole (15.1%), γ-Terpinene (13.7%)	Anesthetized and conscious rats	Hypotension	[60]
Aniba rosaeodora var *amazonica*	Trunk wood	(-)-Linalool (50.6%), (+)-Linalool (49.4%)	Anesthetized rats	Hypotension with bradycardia	[61]
Cymbopogon winterianus	Leaves	Geraniol (40.1%), Citronellal (27.4%), Citronellol (10.5%)	Non-anaesthetized normotensive rats	Hypotension with tachycardia	[62]
Hyptis fruticosa	Leaves	α-Pinene, Caryophyllene, 1,8-Cineole	Non-anesthetized normotensive rats	Hypotension with tachycardia	[63]
Mentha x villosa	Leaves	Piperitenone oxide (95.9%)	DOCA-salt hypertensive rats	↓ MAP without bradycardia	[64]
				Hypotension and ↓ HR	[65]
		Piperitenone oxide (62.3%), γ-Muurolene (16.0%),	Anesthetized rats	Hypotension with bradycardia	[66,67]
		Piperitenone oxide (95.9%)			
		Piperitenone oxide (55.4%), γ-Muurolene (13.1%)	Normotensive Conscious Rats	↓ MAP and HR;	[68]
Pogostemon elsholtzioides	Leaves	Curzene (46.1%)	Anesthetized rats	↓ SBP, DBP, MAP and HR	[69]
Schinus areira	Aerial parts	α-Pinene (13.8%), Limonene (12.8%), Camphene (12.6%)	Non-anesthetized normotensive rats	↓ SBP, DBP and MAP	[70]

DBP – Diastolic blood pressure; DOCA - deoxycorticosterone acetate; HR – Heart rate; MAP – Mean arterial pressure; SBP – Systolic blood pressure.

Accordingly, citronellol caused hypotension associated with tachycardia [71,73]. Geraniol increased the action potential duration and decreased the maximal dp/dt in ventricular cardiomyocytes, decreased the left ventricular pressure, increased PRi and delayed the arrhythmia onset induced by ouabain in guinea pig hearts [74]. Piperitenone oxide decreased the mean arterial pressure and heart rate [66], caused hypotension and bradycardia and induced negative inotropic and chronotropic effect on left and right atrium, respectively [75]. 1,8-Cineole, induced hypotension and bradycardia in a dose-dependent manner [76]. α-Terpineol induced hypotension associated with tachycardia [77] and terpinen-4-ol elicited a decrease in the mean arterial pressure [78]. Other relevant hypotensive monoterpenes are known, such as carvacrol. The first was able to induce hypotension associated with bradycardia [79] and decrease heart rate, mean arterial pressure, as well as systolic and diastolic blood pressures [80].

Some essential oils' activity is related to the presence of active major compounds such as the hypotensive effect reported for *Alpinia zerumbet* that can be associated with the presence of high amounts of terpinen-4-ol and 1,8-cineole [78]. Also, the hypotensive potential of *Mentha x villosa* essential oil is greater in samples with higher amounts of piperitone oxide [66], thus suggesting that this compound is the main active compound in the essential oil.

4.2. Vasorelaxant Effects

The maintenance of the vascular tone is achieved by a balance between vasorelaxant and vasoconstrictor agents, that act on both the vascular smooth muscle cells and the endothelial cells [81]. However, in pathological conditions, such as hypertension, an imbalance between both agents is observed leading to an increase in the vascular tone. This imbalance leads to a pathological condition named endothelial dysfunction. The main vasoactive agents are nitric oxide, prostacyclin, endothelium-derived hyperpolarizing factor, thromboxane A_2, angiotensin II and endothelin-1. Therefore, compounds that help to restore the balance between vasoconstriction and vasorelaxation by either mimicking vasorelaxant agents or antagonizing vasoconstrictor ones, are used for the treatment of endothelial dysfunction. Several terpene-rich essential oils have been widely described as having vasorelaxant activity, mainly in *ex vivo* models, as shown in Table 3.

The mechanisms of action underlying the vasorelaxant effect of essential oils has been described in most cases as an NO-dependent pathway [83,87,95]. Nevertheless, for *Alpinia zerumbet* (fruits) [84] and *Trachyspermum ammi* [96] essential oils, the activity was also reported as dependent on the adenylyl cyclase pathway. Finally, the vasorelaxant activity of *Pogostemon elsholtzioides* [69] and *Artemisia campestris* essential oils was shown to occur though K^+ channel and nitric oxide synthase-independent pathways, being the later dependent on Ca^{2+} channels [85].

For the essential oils, some of their main terpenes have also been pointed out as vasorelaxant agents. Similar to the studies using essential oils, the vasorelaxant potential of terpenes was primarily assessed on phenylephrine, KCl- or $CaCl_2$-induced contractions in aortic rings or mesenteric artery rings. Positive effects were reported for thymol and carvacrol [79,98,99], linalool [72,100,101], citronellol [73], 1,8-cineole [102], citral [95], piperitone oxide [103], terpinen-4-ol [104], and linalyl acetate [105]. Furthermore, carvacrol elicited vasodilation in cerebral arteries by hyperpolarization and decrease in intracellular Ca^{2+} concentration [106]. Also, 1,8-cineole reduced isometric contractions, time to peak and relaxation times as well as tetanic force while increasing the relative potentiation on rat left ventricular papillary muscles [102]. On guinea pig atria, geraniol decreased the contractile force and alleviated the positive ionotropism induced by $CaCl_2$ and BAY K8644 [74]. Other monoterpenes such as carvone [101], α-terpineol [77] and menthol [107,108] have also shown vasorelaxant effects.

In addition to the plethora of studies addressing the vasorelaxant activity of monoterpenes, a few studies have described the same effects for sesquiterpenes. Indeed, caryophyllene oxide was able to induce relaxation in a dose-dependent manner [100]. Also, bisabolol [109,110] showed vasorelaxant effects.

Table 3. Essential oils with vasorelaxant effects

Essential Oil	Part of Plant	Terpenes	Study Model	Cardiovascular Effect	Ref
Alpinia speciosa	Aerial parts	Terpinen-4-ol (38%), 1,8-Cineole (18%), γ-Terpinene (12%)	Rat left atria	↓ Force of contraction in a dose-dependent manner (IC_{50} = 292.2 μg/mL); ↓ Sinus rhythm (IC_{50} = 595.4 μg/mL)	[82]
Alpinia zerumbet	Leaves	1,8-Cineole (33.3%), Terpinen-4-ol (19.4%), p-Cimene (11.4%)	Endothelium-intact aortic rings pre-contracted with Phe	Incomplete relaxation	[83]
	Fruits	β-Phellandrene (16.4%), β-Pinene (15.1%), 1,8-Cineole (11%)	Aortic rings pre-contracted with norepinephrine and KCl	Inhibit the contraction	[84]
Artemisia campestris	Aerial parts	Spathulenol (10.2%)	Endothelium-intact aortic rings contracted with Phe	↓Contraction	[85]
Citrus aurantifolia	Fruits	Limonene (58.4%), β-Pinene (15.4%)	Isolated rabbit aortic rings cultured in high K^+ medium	Relaxation by activating the Ca^{2+} channels	[86]
Citrus aurantium var. *amara*	Commercial	Linalool (23.2%), β-Pinene (9.6%), Limonene (8.54%)	Endothelium-intact aortic rings pre-contracted with $PGF_{2\alpha}$	Vasorelaxation	[87]
Citrus bergamia	Commercial	D-limonene (43.5%), Linalyl acetate (25.5%)	Mouse aortic rings endothelium-intact and -denuded	Inhibited the contraction elicited by $PGF_{2\alpha}$	[88]
Croton argyrophylloides	Aerial parts	Spathulenol (26.7%), Caryophyllene oxide (13.1%), β-Elemene (12.2%)	Endothelium-intact aortic rings and mesenteric arteries pre-contracted with Phe	Vasorelaxation in aortic rings (IC_{50} = 141.1 μg/mL) and mesenteric arteries (IC_{50} = 46.1 μg/mL)	[89]
Croton nepetaefolius	Leaves	1,8-Cineole (25.4%), Bicyclogermacrene (11.1%)	Aortic rings isolated from DOCA-salt-hypertensive rats	Decreased the contractions elicited by Phe	[90]
			Mesenteric vascular bed preparations	Decreased the loss of flow caused by KCl	[91]

Table 3. (Continued)

Essential Oil	Part of Plant	Terpenes	Study Model	Cardiovascular Effect	Ref
	Aerial parts	1,8-Cineole (25.4%)	Endothelium-intact aortic rings	Reduced the contraction evoked by KCl (IC_{50} = 26.7 μg/mL)	[92]
Croton zambesicus	Leaves from different samples	*ent*-Trachyloban-3-one (1.4 – 28.0%), Caryophyllene oxide (2.9% - 25.9%), Longifolene (0.4 – 26.4%)	Endothelium-intact aortic rings	Vasorelaxant activity (IC_{50} = 5.6 – 11.8 μg/mL)	[93]
Cymbopogon winterianus	Leaves	Geraniol (40.1%), Citronellal (27.4%), Citronellol (10.5%)	Mesenteric arteries contracted	Vasorelaxation on rings with (E_{max} = 125%) and without (E_{max} = 117%) endothelium; Vasorelaxation in endothelium-denuded rings precontracted with KCl (E_{max} = 121%).	[62]
Hyptis fruticosa	Leaves	α-Pinene, Caryophyllene, 1,8-Cineole	Endothelium-intact and -denuded rings from the mesenteric artery pre-contracted with Phe	Relaxation (E_{max}= 64% vs 122%, respectively). ↓ Contractions induced by $CaCl_2$ (E_{max} = 12% vs 81%)	[63]
Lippia thymoides	Leaves	β-Caryophyllene (26.3 – 17.2%)	Endothelium-intact and endothelium-denuded rings	Relaxation in endothelium-intact (IC_{50} = 305 – 544 μg/mL) and endothelium-denuded (IC_{50} = 150 – 283 μg/mL) rings	[94]
Mentha x villosa	Leaves	Piperitenone oxide (95.9%)	Isolated atrial preparations aortic rings	Dose-dependent negative chronotropic (IC_{50} = 229 μg/mL) and ionotropic (IC_{50} = 120 μg/mL) effects. Relaxation on aortic contracted with Phe- (IC_{50} = 255 μg/mL), $PGF_{2\alpha}$-induced (IC_{50} = 174 μg/mL) and KCl (IC_{50} = 165 μg/mL).	[65]
			Isolated aortic rings contracted by KCl	Relaxation (IC_{50} = 61 μg/mL *vs* 109 μg/mL for endothelium-intact and denuded rings, respectively)	[64]

Essential Oil	Part of Plant	Terpenes	Study Model	Cardiovascular Effect	Ref
Pectis brevipedunculata	Aerial parts	Citral [Neral (32.7%) + Geranial 49.2%)]	Phe-contracted aortic rings	Vasorelaxation on endothelium-intact (IC_{50} = 0.044%) and endothelium-denuded (IC_{50} = 0.093%) rings.	[95]
Pogostemon elsholtzioides	Leaves	Curzene (46.1%)	Aortic rings pre-contracted with Phe	Relaxation	[69]
Schinus areira	Aerial parts	α-Pinene (13.8%), Limonene (12.8%), Camphene (12.6%), β-Caryophyllene (11.9%)	*ex vivo* model of rabbit hearts	Inhibit the cardiac contractility induced by norepinephrine	[70]
Trachyspermum ammi	Seeds	Thymol (38.1%), Limonene (33.3%), p-Cymene (23.1%)	Aortic rings	↓ Contraction of aortic rings induced by Phe (IC_{50} = 54.4 μg/mL), KCl (IC_{50} = 49 μg/mL) in the presence (IC_{50} = 46.6 μg/mL) and absence (IC_{50} = 45.2 μg/mL) of endothelium	[96]
Xylopia langsdorfiana	Leaves	Germacrene D (22.9%), trans-β-Guaiene (22.6%), β-Caryophyllene (15.7%)	Isolated aortic rings contracted with Phe	Weakly inhibits the contraction	[97]

$CaCl_2$ – Calcium chloride; DOCA - Deoxycorticosterone acetate; E_{max} – Ventricular end-systolic maximum elastance; IC_{50} – Concentration needed to achieve 50% of relaxation; KCl – Potassium chloride; $PGF_{2\alpha}$ – Prostaglandin $F_{2\alpha}$; Phe – Phenylephrine

In what concerns composition-activity relations, α-pinene has been reported as a smooth muscle relaxant [111] and may be responsible for the vasorelaxant effect of *Hyptis fruticosa* essential oils that present high amounts of this compound [63]. Also the vasorelaxant activity of *Citrus aurantium* var. *amara* can be explained by the presence of linalool, since this compound also elicits a relaxant activity dependent of the nitric oxide/cyclic guanosine monophosphate pathway [87]. The essential oil of bergamot (*Citrus bergamia*) also elicited a vasorelaxant effect that can be partially explained by the presence of linalool and linalyl acetate [88]. *Croton nepetaefolius* essential oil's vasorelaxant activity might be explained by the presence of 1,8-cineole and α-terpineol [92]. Nevertheless, the activity described by *Alpinia zerumbet* cannot be fully attributed to the presence of 1,8-cineole, because this compound elicits a full relaxation whereas the essential oil only elicited a partial one [83], thus suggesting an antagonistic effect of other compounds present in the mixture.

4.3. Ion Channel Modulator Effects

Calcium is relevant in several physiological and pathological situations on different organ systems [112]. In the cardiovascular system, calcium is a messenger in muscle contractility as well as in platelet aggregation. In addition, in some pathologies, the intracellular calcium release during diastole is impaired, thus decreasing the relaxation needed for the correct functioning of the heart [113]. Furthermore, high extracellular concentrations of this ion are associated with increased risk of CVDs. Therefore, compounds that are able to maintain an adequate intracellular amount of calcium are important for the correct functioning of the heart. In fact, essential oils rich in terpenes have been described as efficient promoters of calcium intake and release, as shown in Table 4.

Regarding these essential oils, only in the case of neroli (*Citrus aurantium* var. *amara*), a potential mechanism of action has been shown suggesting that the modulation of intracellular Ca^{2+} concentration occurs through the inhibition of channel-mediated extracellular Ca^{2+} influx and store-operated Ca^{2+} release mediated by the Ryanodine Receptors signalling pathway [87].

Table 4. Essential oils with ion channel modulator effects

Essential Oil	Part of Plant	Terpenes	Study Model	Cardiovascular Effect	Ref
Alpinia speciosa	Aerial parts	Terpinen-4-ol (38%), 1,8-Cineole (18%)	Whole-cell clamps	↓ Intercellular calcium (32.6% at 25 µg/mL *vs* 89.3% at 250 µg/mL)	[82]
Citrus aurantium var. *amara*	Commercial	Linalool (23.21%), β-pinene (9.59%), Limonene (8.54%)	Smooth muscle cells	Relaxation caused by modulation of intracellular Ca^{2+}	[87]
Citrus bergamia	Fruits	Limonene (43.5%), Linalyl acetate (25.2%)	Mouse endothelial and vascular smooth muscle cells	Endothelial cells: Transient ↑ intracellular Ca^{2+} followed by a decrease. Vascular smooth muscle cells: sustained ↑ intracellular calcium	[114]
Nardostachys jatamansi	Roots and rhizomes	Calarene (38%), β-Maaliene (7.9%), Valerena-4,1(11)-diene (6.6%)	HUVEC	↑ Intracellular Ca^{2+}	[115]

HUVEC - Human umbilical vein endothelial cells

Once again, this cardioprotective effect, may be related to the presence of active compounds, namely monoterpenes and sesquiterpenes. Indeed, some studies have shown the potential of thymol and carvacrol. For example, on canine myoblasts, these compounds inhibited the L-type Ca^{2+} current in a dose-dependent manner [116]. On isolated human and canine ventricular cardiomyoblasts, thymol supressed the activity of Ca^{2+} and K^+ channels [117] and triggered the release of Ca^{2+} from the sarcoplasmic reticulum while blocking the activity of Ca^{2+} pump [118]. Also, carvacrol inhibited the Ca^{2+} influx by L-type Ca^{2+}-channels [79] and increased the intracellular concentration of this ion in a dose-dependent manner in cerebral artery endothelial cells [106]. 1,8-Cineole was able to decrease the contractility of left ventricular papillary muscles by reducing the sarcolemmal Ca^{2+} influx [102]. Regarding sesquiterpenes, and similarly to what was described for vasorelaxation effects, studies addressing terpene's ion channel modulating potential are scarce. For example, in guinea pig ventricle isolated cardiomyocytes, it was shown that β-caryophyllene oxide inhibited both Ca^{2+} and K^{+} currents in a dose-dependent manner, thus pointing out a weak negative ionotropic effect [119].

The ion channel modulation activity of some essential oils may be due to the presence of active terpenes. For example, the effect of *Alpina speciosa* is linked to the presence of 1,8-cineole. Nevertheless, terpinene-4-ol [120] and γ-terpinene [121] have also caused relaxation in a non-cardiac muscle in a Ca^{2+} dependent manner, thus suggesting that these compounds are also responsible for the activity of the whole essential oil [82]. Linalool was able to decrease the contraction in mesenteric arteries elicited by extracellular Ca^{2+} [72]. This activity might explain the effect of *Citrus aurantium* var. *amara* essential oil since the effect on contractions is modulated, partially, by the inhibition of Ca^{2+} influx [87]. Linalyl acetate blocked Ca^{2+} influx in both endothelial and smooth muscle cells [114]. However, the activity of this compound is lower than that of *Citrus bergamia* essential oil, thus suggesting that other compounds, including minor ones might have additional effects and promote a stronger mobilization of intracellular Ca^{2+} [114].

4.4. Antiplatelet Aggregation Effects

Platelet aggregation is fundamental in physiological conditions to prevent the loss of blood during haemorrhaging. However, in pathological conditions, platelets can hyperaggregate leading to the formation of thrombus [122]. This hyperaggregability is caused by an overproduction of pro-aggregatory factors and/or a subproduction of anti-aggregatory agents. Moreover, several risk factors for CVDs can induce platelets hyperactivation. To avoid this, antiplatelet drugs are used. Despite the widespread use of aspirin, clopidogrel and glycoprotein IIb/IIIa inhibitors, the response of patients to therapy shows great variability due to gene polymorphisms as well as clinical and/or environmental factors [123]. Therefore, new antiplatelet aggregation agents are required to improve the overall response to therapy. In fact, some plant volatile extracts have been described as possessing these effects, as summarized in Table 5.

Table 5. Essential oils with antiplatelet aggregation effects

Essential Oil	Part of Plant	Terpenes	Study Model	Cardiovascular Effect	Ref
Artemisia campestris	Leaves and flowers	Spathulenol (10.2%)	Wistar rats and albino mice	↓ Aggregation induced by thrombin (49.73% at 1 mg/mL) and ADP (48.20% at 1 mg/mL)	[85]
Monarda didyma	Stems	Geraniol (89.5%)	Guinea pig and rat plasma	↓ AA-induced platelet aggregation (IC_{50} = 13 µg/mL)	[124]
Ocimum basilicum	Stems	Linalool (49.9%)	Guinea pig and rat plasma	↓ AA-induced platelet aggregation (IC_{50} = 22 µg/mL)	[124]
Origanum vulgaris	Stems	Carvacrol (54.4%), Thymol (14.3%)	Guinea pig and rat plasma	↓ AA-induced platelet aggregation (IC_{50} = 1.9 µg/mL)	[124]
Syringa pinnatifolia var. *alashanensis*	Stems	α-Cadinol (19.9%), α-Muurolol (18.5%)	Primary cultured rat neonatal myocytes	↓ ADP-induced platelet aggregation	[125]
Thymus vulgaris	Stems	p-Cymene (15.3%)	Guinea pig and rat plasma	↓ AA-induced platelet aggregation (IC_{50} = 4.7 µg/mL)	[124]

AA – Arachidonic acid; ADP – Adenosine diphosphate; IC_{50} – Concentration required to achieve 50% inhibition of platelet aggregation

To the authors knowledge, studies reporting the mechanism of action underlying the essential oils antiplatelet aggregation effect are missing. Also, the activity of isolated terpenes found in essential oils as well as correlations between the effect of the total extract and the isolated compounds are unknown.

4.5. Anti-Dyslipidaemic Effects

Lipid metabolism can be disturbed causing changes in lipoprotein function and/or levels [126]. Although these alterations are not a direct risk for CVDs, their interaction with different cardiovascular risk factors can lead to atherosclerosis. Although dyslipidaemia includes a wide spectrum of lipids, the most widely studied include the levels of total cholesterol and low-density lipoprotein-cholesterol (LDL-C) since these can be modulated by lifestyle changes and drug therapy. Nevertheless, other lipids can also be altered in dyslipidaemia, such as very low-density lipoprotein (VLDL-C), triglycerides and high-density lipoprotein-cholesterol (HDL-C). Therefore, compounds that modulate lipid levels either by inhibiting their absorption in the gut, such as phytosterols [127] or by modulating the activity of lipid metabolism enzymes, such as 3-hydroxy-3-methyl-glutaryl-CoA, acetyl CoA acetyltransferase and sterol regulatory element-binding protein [128,129] are good candidates for lipid-lowering agents. Indeed, some essential oils are known to have these properties as described in Table 6.

The study on the anti-dyslipidaemic effects of turmeric (*Curcuma longa*) essential oils pointed out a possible mechanism of action by suggesting that the oil was able to ameliorate the oxidative stress and liver dysfunction elicited by the diet, being this effect mediated by the modulation of peroxisome proliferator-activated receptor α, liver X receptor α and associated genes involved in lipid metabolism and transport [130].

Regarding isolated terpenes, some studies have focused on the effect of monoterpenes. For example, thymol was able to decrease haemoglobin A1c, blood glucose, insulin, triglycerides, total cholesterol, free fatty acids, LDL-C and leptin while increasing the levels of HDL-C and adiponectin, in mice with type 2 diabetes mellitus induced by high fat diet (HFD) [133]. Its isomer, carvacrol, decreased the total cholesterol, total triglycerides, free fatty acids and phospholipids in the plasma, liver, heart and kidney in diabetic mice fed a HFD [134]. Moreover, the presence of carvacrol modulated the cholesterol type in the plasma of HFD-fed mice with an increase in HDL-C and a decrease on both VLDL-C and LDL-C. The treatment also changed histopathological features, namely lower fat accumulation in adipocytes, fewer lipid droplets on the liver and decreased the expression of Tumor necrosis factor-α and Interleukin 6 [134].

Table 6. Essential oils with anti-dyslipidaemic effects

Essential Oil	Part of Plant	Terpenes	Study Model	Cardiovascular Effect	Ref
Curcuma longa	Dried rhizomes	ar-Turmerone (31.7%), β-Turmerone (14.3%), α-Turmerone (11.5%)	Golden Syrian hamsters consuming a high cholesterol diet	↓ TC, LDL-C and TG ↑HDL-C in plasma (100 and 300 mg/kg); ↓ hepatic TC, free cholesterol and cholesteryl ester.	[130]
Pinus koraiensis	Leaves	Camphene (21.1%), D-Limonene (21.0%), α-Pinene (16.7%)	HepG2 cells	↑ mRNA and protein levels of LDL receptor; ↓ mRNA levels SREBP-1c, SREBP-2, HMG-CoA reductase, FAS and GPAT; ↓ activity of hACAT 1 and 2; ↓ oxidation of LDL	[131]
Plantago asiatica	Leaves	Linalool (82.5%)	HepG2 cells	↑ LDL receptor ↓ HMG-CoA reductase and LDL oxidation	[132]
			C57BL/6 mice	↓ TC, TG ↓ mRNA and protein levels of HMG-CoA reductase; ↑ mRNA of LDL receptor	

FAS – Fatty acid synthase; GPAT – Glycerol-3-phosphate acyltransferase; hACAT – Human acetyl CoA acetyltransferase; HDL-C – High density lipoprotein-cholesterol; HepG2 - liver hepatocellular cells; HMG-CoA - 3-Hydroxy-3-methyl-glutaryl-CoA; LDL – Low density lipoprotein; LDL-C – Low density lipoprotein-cholesterol; SREBP – Sterol regulatory element-binding protein; TC – Total cholesterol; TG – Total triglycerides

Geraniol was able to decrease serum cholesterol and total triglycerides as well as hepatic free cholesterol, esterified cholesterol and total triglycerides on NIH *nu/nu* mice [135]. Furthermore, total liver lipids, nonsaponifiable lipids and fatty acids synthesis were decreased after geraniol supplementation. Both the mRNA and protein levels of 3-hydroxy-3-methylglutaryl-coenzyme A (HMG-CoA) reductase were increased as well as the specific activity after the treatment. In addition, the mRNA levels of LDL and VLDL receptors were increased whereas acetyl-CoA carboxylase 1-receptor was decreased. On atherogenic diet-fed Syrian hamsters, geraniol decreased total cholesterol, total triglycerides, free fatty acids and phospholipids [136]. Furthermore, it was able to decrease LDL and VLDL while increasing HDL. In addition, the treatment decreased C reactive protein and HMG-CoA reductase activity. Geraniol increased the activity of lipoprotein lipase and lecithin–cholesterol acyltransferase and alleviated cardiac hypertrophy. The administration of camphene on hyperlipidaemic rats caused a decrease in total cholesterol, LDL-C and total triglycerides [137].

Some studies have correlated the anti-dyslipidaemic effect of some essential oils with their isolated compounds. It was shown that linalool could affect LDL metabolism by decreasing the oxidation of LDL as well as increasing the affinity to LDL receptor [138] and therefore, seems to be responsible for *Plantago asiatica* essential oil's effect [132]. Also, the antidyslipidaemic activity of *Pinus koraiensis* essential oil might be partially explained by the antidyslipidaemic activity of camphene, its major compound, although the authors also suggest a synergistic effect with other compounds [131].

4.6. Protection against Myocardial Infarction Injury

Myocardial infarction (MI) is characterized by heart muscle injury, which is consistent with an increase in the troponin levels to at least one level above the 99th percentile, resulting from ischemia induced necrosis of cardiomyocytes [139]. Furthermore, MI is characterized by chest discomfort and ST-deviation. Thus, therapies that modulate risk factors as well as thrombus formation are rising due to their importance in the management of MI. One study has reported beneficial effects of an essential oil, as summarized in Table 7.

Table 7. Essential oils with protective effect against myocardial infarction

Essential Oil	Part of Plant	Terpenes	Study Model	Cardiovascular Effect	Ref
Syringa pinnatifolia	Stems	α-Cadinol (19.9%), τ-Muurolol (18.5%)	Wistar rats, Kunming mice and primary cultured rat neonatal myocytes	↓ Deviation on ST-segment, ↓ LDH, CK and TnT; ↑ SOD activity	[125]

CK – Creatine kinase; LDH – Lactate dehydrogenase; SOD – Superoxide dismutase; TnT – Troponin T

Regarding isolated terpenes, it is known that the oral administration of farnesol, an acyclic sesquiterpene alcohol decreased infarct size after ischemia/reperfusion (I/R) events and prevented cell death in isolated cardiomyocytes, after simulated I/R [140].

4.7. Other Beneficial Effects on the Cardiovascular System

Besides the effects described above, terpene-rich essential oils have also been reported for other positive effects related to the cardiovascular system, such as inducers of cell survival under stressful conditions, as shown in Table 8.

Regarding isolated terpenes, carvacrol was able to decrease rat aortic smooth muscle cells migration and proliferation associated with platelet-derived growth factor (PDGF) in a dose-dependent manner [144]. Furthermore, it decreased ROS production by decreasing both NADPH oxidase 1 (NOX1) activity and expression. In addition, the phosphorylation of Extracellularly-Regulated Kinase-1 and -2 (ERK1/2) and p38 mitogen-activated protein kinases (p38 MAPK) evoked by PDGF was inhibited in the presence of this compound. This compound also inhibited the outgrowth of aortic sprouts as well as neointima formation. On the other hand, inhalation of both linalool isomers, (-)- and (+)-linalool, decreased SBP, DBP and heart rate on healthy human subjects [145]. Borneol increased cell viability on hypoxia/reoxygenation-stimulated cardiomyocytes [146]. Thujone diminished total cholesterol and triglyceride levels [147] and β-caryophyllene reduced blood glucose and increased the levels of insulin as well as the activity of antioxidant enzymes and agents [148], thus promoting an antidiabetic effect.

Table 8. Essential oils with other protective effects

Essential Oil	Part of Plant	Terpenes	Study Model	Cardiovascular Effect	Ref
Alpinia zerumbet	Fruits	β-Phellandrene (16.4%), β-Pinene (15.1%), 1,8-Cineole (11%)	HUVEC HAEC treated with oxLDL	↑ cell viability in oxLDL-induced injury in HUVEC; ↓ LDH release (328.68 vs 555.15 U/L) and MDA levels; ↑ GSH contents. ↑ SOD, CAT GSH-Px activity	[141]
				↑ cell viability ↓ LDH release; ↑ MMP; ↓ ROS production; ↑	[142]
				NO production; ↑ mRNA and protein levels of Akt/p-Akt, eNOS and sGC; ↓iNOS levels	
Ocimum basilicum (cultivar Genovese)	Seeds	Linalool (38.3%)	Primary culture cardiomyocytes treated with H_2O_2	↑ cell proliferation	[143]
Ocimum basilicum (cultivar Lettuce leaf)	Seeds	Linalool (36%)	Primary culture cardiomyocytes treated with H_2O_2	↑ cell proliferation	[143]
Ocimum basilicum (cultivar Purple)	Seeds	Linalool (47.5%)	Primary culture cardiomyocytes treated with H_2O_2	↑ cell proliferation	[143]
Syringa pinnatifolia	Stems	α-Cadinol (19.9%), τ-Muurolol (18.5%)	Wistar rats, Kunming mice and primary cultured rat neonatal myocytes	↑ survivability of rats under hypoxic conditions; ↓ H_2O_2-induced cell death	[125]

CAT – Catalase; eNOS – Endothelial nitric oxide synthase; GSH – Glutathione; GSH-Px – Glutathione peroxidase; H_2O_2 – Hydrogen peroxide; HAEC - Human aortic endothelial cells; HUVEC – Human umbilical vessel endothelial cells; iNOS – Inducible nitric oxide synthase; LDH - Lactate dehydrogenase; MDA- Malondialdehyde; MMP – mitochondrial membrane potential; NO – Nitric oxide; oxLDL – oxidized low density lipoprotein; sGC – Soluble guanylate cyclase; SOD – Superoxide dismutase.

CONCLUSION

The research conducted on the cardioprotective effects of terpene-rich essential oils and isolated terpenes, has highlighted the interest of the scientific and medical community on plant-based therapies. The hypotensive and vasorelaxant potential of these extracts has been been widely reported however, only a few studies refer the ion channel modulating, antiplatelet aggregation and anti-dyslipidaemic effects. Interestingly, many studies focus on the cardioprotective potential of phenolic monoterpenes such as thymol and carvacrol, but lack to evaluate the effect of extracts rich in these compounds. Nevertheless, some studies have identified other terpenes responsible for the cardioprotective potential of some essential oils, namely 1,8-cineole, terpinen-4-ol and piperitenone oxide, whereas other studies point out possible synergistic and antagonistic effects among the extract's compounds.

Overall, these studies are very significant since they corroborate many traditional uses of aromatic and medicinal plants, by validating their beneficial effects on relevant pathophysiological models. Indeed, *A. zerumbet* fruits have been extensively used for the treatment of CVDs in China, Brazil and Japan [60,149–151]. Also, *Citrus aurantifolia* and *C. aurantium*, are used traditionally for the treatment of hypertension in Ethiopia [152] and for diabetes in Guatemala [153], respectively. *Croton* species are prescribed by traditional medicine practitioners for the treatment of type 2 diabetes [153], and to ameliorate dyslipidaemia and diabetes in Brazil [154], while turmeric (*Curcuma longa*) is widely used in Ayurvedic medicine for the treatment of diabetes [155], as well as for the management of hypertension and vascular disorders in western African folk medicine [156]. Moreover, some of these studies identify new putative active compounds, that are particularly interesting for the development of novel therapeutic approachcs. Nevertheless, many of these studies lack mechanistic evaluations and more comparative studies between extracts and isolated compounds are needed. Moreover, clinical trials with herbal drugs are challenging since they lack stringent bylaws and regulations. Indeed, appropriate requirements should be established to assess extracts safety and efficacy in humans, since quality control of herbal medicines is very difficult.

REFERENCES

[1] WHO | *Cardiovascular diseases* (CVDs). WHO 2017.

[2] WHO | *World Heart Day* 2017. WHO 2017.

[3] Walden R, Tomlinson B. *Cardiovascular disease.* 2nd Editio. CRC Press/Taylor & Francis; 2011.

[4] Cordell G. Changing strategies in natural products chemistry. *Phytochemistry* 1995;40:1585–612. doi:10.1016/0031-9422(95) 00444-C.

[5] Tuso P, Stoll SR, Li WW. A plant-based diet, atherogenesis, and coronary artery disease prevention. *Perm J* 2015;19:62–7. doi:10.7812/TPP/14-036.

[6] Ornish D. Intensive lifestyle changes for reversal of coronary heart disease. *JAMA* 1998;280:2001. doi:10.1001/jama.280.23.2001.

[7] Wachtel-Galor S, Benzie IFF. Herbal Medicine: An introduction to its history, usage, regulation, current trends, and research needs. In: Wachtel-Galor S, Benzie IFF, editors. *Herb Med Biomol Clin Asp.* 2nd Editio, CRC Press/Taylor & Francis; 2011.

[8] Jenke-Kodama H, Müller R, Dittmann E. Evolutionary mechanisms underlying secondary metabolite diversity. *Prog Drug Res* 2008;65:119, 121–40.

[9] Alves-Silva JM, Zuzarte M, Marques C, Salgueiro L, Girao H. Protective effects of terpenes on the cardiovascular ystem: current advances and future perspectives. *Curr Med Chem* 2016;23:4559–600.

[10] Townsend N, Wilson L, Bhatnagar P, Wickramasinghe K, Rayner M, Nichols M. Cardiovascular disease in Europe: epidemiological update 2016. *Eur Heart J* 2016;37:3232–45. doi:10.1093/eurheartj /ehw334.

[11] Nichols M, Townsend N, Scarborough P, Rayner M. Cardiovascular disease in Europe: epidemiological update. *Eur Heart J* 2013;34:3028–34. doi:10.1093/eurheartj/eht356.

[12] Mendis S, Puska P, Norrving B, World Health Organization., World Heart Federation., World Stroke Organization. *Global atlas on cardiovascular disease prevention and control.* World Health Organization in collaboration with the World Heart Federation and the World Stroke Organization; 2011.

[13] Vos T, Flaxman AD, Naghavi M, Lozano R, Michaud C, Ezzati M, et al. Years lived with disability (YLDs) for 1160 sequelae of 289 diseases and injuries 1990–2010: a systematic analysis for the global burden of disease study 2010. *Lancet* 2012;380:2163–96. doi:10.1016/S0140-6736(12)61729-2.

[14] Stewart J, Manmathan G, Wilkinson P. Primary prevention of cardiovascular disease: A review of contemporary guidance and literature. *JRSM Cardiovasc Dis* 2017;6:204800401668721. doi:10.1177/2048004016687211.

[15] Scott J. Pathophysiology and biochemistry of cardiovascular disease. *Curr Opin Genet Dev* 2004;14:271–9. doi:10.1016/j.gde.2004.04.012.

[16] Montalescot G, Sechtem U, Achenbach S, Andreotti F, Arden C, Budaj A, et al. 2013 ESC guidelines on the management of stable coronary artery disease. *Eur Heart J* 2013;34:2949–3003. doi:10.1093/eurheartj/eht296.

[17] Aboyans V, Ricco J-B, Bartelink M-LEL, Björck M, Brodmann M, Cohnert T, et al. 2017 ESC Guidelines on the diagnosis and treatment of peripheral arterial diseases, in collaboration with the European Society for Vascular Surgery (ESVS). *Eur Heart J* 2017. doi:10.1093/eurheartj/ehx095.

[18] Seshadri S, Debette S. *Risk factors for cerebrovascular disease and stroke.* Oxford University Press; 2016.

[19] Watkins DA, Johnson CO, Colquhoun SM, Karthikeyan G, Beaton A, Bukhman G, et al. Global, regional, and national burden of

rheumatic heart disease, 1990–2015. *N Engl J Med* 2017;377:713–22. doi:10.1056/NEJMoa1603693.

[20] Baumgartner H, Bonhoeffer P, De Groot NMS, de Haan F, Deanfield JE, Galie N, et al. ESC Guidelines for the management of grown-up congenital heart disease (new version 2010): The Task Force on the Management of Grown-up Congenital Heart Disease of the European Society of Cardiology (ESC). *Eur Heart J* 2010;31:2915–57. doi:10.1093/eurheartj/ehq249.

[21] Kelleher AA. Adult congenital heart disease (grown-up congenital heart disease). *Contin Educ Anaesth Crit Care Pain* 2012;12:28–32. doi:10.1093/bjaceaccp/mkr045.

[22] Mazzolai L, Aboyans V, Ageno W, Agnelli G, Alatri A, Bauersachs R, et al. Diagnosis and management of acute deep vein thrombosis: a joint consensus document from the European society of cardiology working groups of aorta and peripheral vascular diseases and pulmonary circulation and right ventricular function. *Eur Heart J* 2017. doi:10.1093/eurheartj/ehx003.

[23] Buttar HS, Li T, Ravi N. Prevention of cardiovascular diseases: Role of exercise, dietary interventions, obesity and smoking cessation. *Exp Clin Cardiol* 2005;10:229–49.

[24] Harvey RE, Coffman KE, Miller VM. Women-specific factors to consider in risk, diagnosis and treatment of cardiovascular disease. *Women's Heal* 2015;11:239–57. doi:10.2217/WHE.14.64.

[25] Leigh JA, Alvarez M, Rodriguez CJ. Ethnic minorities and coronary heart disease: an update and future directions. *Curr Atheroscler Rep* 2016;18:9. doi:10.1007/s11883-016-0559-4.

[26] Ng C-Y, Leong X-F, Masbah N, Adam SK, Kamisah Y, Jaarin K. Heated vegetable oils and cardiovascular disease risk factors. *Vascul Pharmacol* 2014;61:1–9. doi:10.1016/j.vph.2014.02.004.

[27] Teo K, Lear S, Islam S, Mony P, Dehghan M, Li W, et al. Prevalence of a healthy lifestyle among individuals with cardiovascular disease in high-, middle- and low-income countries. *JAMA* 2013;309:1613. doi:10.1001/jama.2013.3519.

[28] Perk J, De Backer G, Gohlke H, Graham I, Reiner Z, Verschuren M, et al. European Guidelines on cardiovascular disease prevention in clinical practice (version 2012). The Fifth Joint Task Force of the European Society of Cardiology and Other Societies on Cardiovascular Disease Prevention in Clinical Practice (constituted by re. *Eur Heart J* 2012;33:1635–701. doi:10.1093/eurheartj/ ehs092.

[29] Kromhout D, Menotti A, Kesteloot H, Sans S. Prevention of coronary heart disease by diet and lifestyle: evidence from prospective cross-cultural, cohort, and intervention studies. *Circulation* 2002;105:893–8.

[30] Blood Pressure Lowering Treatment Trialists' Collaboration, Turnbull F, Neal B, Ninomiya T, Algert C, Arima H, et al. Effects of different regimens to lower blood pressure on major cardiovascular events in older and younger adults: meta-analysis of randomised trials. *BMJ* 2008;336:1121–3. doi:10.1136/bmj.39548. 738368.BE.

[31] Costa APR, de Paula RCS, Carvalho GF, Araújo JP, Andrade JM, de Almeida OLR, et al. High sodium intake adversely affects oxidative-inflammatory response, cardiac remodelling and mortality after myocardial infarction. *Atherosclerosis* 2012;222:284–91. doi:10.1016/j.atherosclerosis.2012.02.037.

[32] Hassan M, Aguib Y, Yacoub M. Molecular mechanisms of cardiovascular benefits of exercise: Running for cover from heart disease. *Glob Cardiol Sci Pract* 2016;2016. doi:10.21542/ gcsp.2016.3.

[33] Booth FW, Roberts CK, Laye MJ. Lack of exercise is a major cause of chronic diseases. *Compr Physiol*, Hoboken, NJ, USA: John Wiley & Sons, Inc.; 2012. doi:10.1002/cphy.c110025.

[34] U.S. Department of Health and Human Services. *How tobacco smoke causes disease: the biology and behavioral basis for smoking-attributable disease*. 2010. doi:10.1037/e590462011-001.

[35] Zakhari S. Alcohol and the cardiovascular system: molecular mechanisms for beneficial and harmful action. *Alcohol Health Res World* 1997;21:21–9. doi:10.1016/j.jacc.2005.02.075.
[36] Baumgartner H, Falk V, Bax JJ, De Bonis M, Hamm C, Holm PJ, et al. 2017 ESC/EACTS Guidelines for the management of valvular heart disease. *Eur Heart J* 2017;38:2739–91. doi:10.1093/eurheartj/ehx391.
[37] Knoops KTB, de Groot LCPGM, Kromhout D, Perrin A-E, Moreiras-Varela O, Menotti A, et al. Mediterranean diet, lifestyle factors, and 10-year mortality in elderly European men and women: the HALE project. *JAMA* 2004;292:1433–9. doi:10.1001/jama.292.12.1433.
[38] Esposito K, Marfella R, Ciotola M, Di Palo C, Giugliano F, Giugliano G, et al. Effect of a mediterranean-style diet on endothelial dysfunction and markers of vascular inflammation in the metabolic syndrome: a randomized trial. *JAMA* 2004;292:1440–6. doi:10.1001/jama.292.12.1440.
[39] Anand SS, Bosch J, Eikelboom JW, Connolly SJ, Diaz R, Widimsky P, et al. Rivaroxaban with or without aspirin in patients with stable peripheral or carotid artery disease: an international, randomised, double-blind, placebo-controlled trial. *Lancet* 2018;391:219–29. doi:10.1016/S0140-6736(17)32409-1.
[40] Faulx MD, Francis GS. Adverse Drug Reactions in Patients with Cardiovascular Disease. *Curr Probl Cardiol* 2008;33:703–68. doi:10.1016/j.cpcardiol.2008.08.002.
[41] Council of Europe. *European Pharmacopoeia* 7th Edition. Strasbourg, France: Directorate for the quality of medicines & healthcare of the Council of Europe; 2010.
[42] Svoboda KP, Svoboda TG, Syred PM. *Secretory structures of aromatic and medicinal plants: A review and atlas of micrographs*. Microscopix; 2000.
[43] Lahlou M. Methods to study the phytochemistry and bioactivity of essential oils. *Phytother Res* 2004;18:435–48. doi:10.1002/ptr.1465.

[44] Rubiolo P, Sgorbini B, Liberto E, Cordero C, Bicchi C. Essential oils and volatiles: sample preparation and analysis. A review. *Flavour Fragr J* 2010;25:282–90. doi:10.1002/ffj.1984.

[45] Adams RP. *Identification of essential oil components by gas chromatography/mass spectroscopy*. Illinois, USA: Allured Publishing Corporation, Carol Stream; 2007.

[46] Joulain D, König WA. *The atlas of spectral data of sesquiterpene hydrocarbons*. Hamburg: E.B.-Verlag; 1998.

[47] Ferreira MJ. P, Costantin MB, Rodrigues G V, Limberger R, Henriques AT, Kato MJ, et al. Computer-aided method for identification of components in essential oils by 13C NMR spectroscopy. *Anal Chim Acta* 2001;447:125–34. doi:10.1016/S0003-2670(01)01204-1.

[48] Figueiredo AC, Barroso JG, Pedro LG, Scheffer JJC. Factors affecting secondary metabolite production in plants: volatile components and essential oils. *Flavour Fragr J* 2008;23:213–26. doi:10.1002/ffj.1875.

[49] Taiz L, Zeiger E. *Plant physiology*. 5th ed. Sinauer Associates; 2010. doi:10.1104/pp.900074.

[50] Bakkali F, Averbeck S, Averbeck D, Idaomar M. Biological effects of essential oils – A review. *Food Chem Toxicol* 2008;46:446–75. doi:10.1016/j.fct.2007.09.106.

[51] Hüsnü K, Başer C, Demirci F. Chemistry of essential oils. In: Berger RG, editor. *Flavours and Fragrances*, Berlin, Heidelberg: Springer Berlin Heidelberg; 2007, p. 43–86. doi:10.1007/978-3-540-49339-6_4.

[52] Zuzarte M, Salgueiro L. Essential oils chemistry. *Bioact Essent Oils Cancer, Cham*: Springer International Publishing; 2015, p. 19–61. doi:10.1007/978-3-319-19144-7_2.

[53] Bouwmeester HJ. Engineering the essence of plants. *Nat Biotechnol* 2006;24:1359–61. doi:10.1038/nbt1106-1359.

[54] Buchbauer G. The detailed analysis of essential oils leads to the understanding of their properties. *Perfumer&Flavorist* 2000;25:64–7.

[55] HMPC/EMA. *Herbal medicine for human use* n.d.

[56] Rastogi S, Pandey MM, Rawat AKS. Traditional herbs: a remedy for cardiovascular disorders. *Phytomedicine* 2016;23:1082–9. doi:10.1016/j.phymed.2015.10.012.

[57] Mancia G, Fagard R, Narkiewicz K, Redon J, Zanchetti A, Böhm M, et al. 2013 ESH/ESC Guidelines for the management of arterial hypertension. *Eur Heart J* 2013;34:2159–219. doi:10.1093/eurheartj/eht151.

[58] Conroy R. Estimation of ten-year risk of fatal cardiovascular disease in Europe: the SCORE project. *Eur Heart J* 2003;24:987–1003. doi:10.1016/S0195-668X(03)00114-3.

[59] Trinder Y. Common and less common adverse effects of antihypertensives: A general practitioner's perspective. *South African Fam Pract* 2012;54:S31–2.

[60] Lahlou S, Galindo CAB, Leal-Cardoso JH, Fonteles MC, Duarte GP. Cardiovascular effects of the essential oil of *Alpinia zerumbet* leaves and its main constituent, terpinen-4-ol, in rats: role of the autonomic nervous system. *Planta Med* 2002;68:1097–102. doi:10.1055/s-2002-36336.

[61] de Siqueira RJ, Rodrigues KMS, da Silva MTB, Correia Junior CAB, Duarte GP, Magalhães PJC, et al. Linalool-rich rosewood oil induces vago-vagal bradycardic and depressor reflex in rats. *Phytother Res* 2014;28:42–8. doi:10.1002/ptr.4953.

[62] de Menezes IAC, Moreira IJA, de Paula JWA, Blank AF, Antoniolli AR, Quintans-Júnior LJ, et al. Cardiovascular effects induced by *Cymbopogon winterianus* essential oil in rats: involvement of calcium channels and vagal pathway. *J Pharm Pharmacol* 2010;62:215–21. doi:10.1211/jpp.62.02.0009.

[63] Santos MRV, Carvalho AA, Medeiros IA, Alves PB, Marchioro M, Antoniolli AR. Cardiovascular effects of *Hyptis fruticosa* essential oil in rats. *Fitoterapia* 2007;78:186–91. doi:10.1016/j.fitote.2006.11.009.

[64] Lahlou S, Carneiro-Leão RFL, Leal-Cardoso JH. Cardiovascular effects of the essential oil of *Mentha* x *villosa* in DOCA-salt-hypertensive rats. *Phytomedicine* 2002;9:715–20. doi:10.1078/094471102321621313.

[65] Guedes DN, Silva DF, Barbosa-Filho JM, de Medeiros IA. Endothelium-dependent hypotensive and vasorelaxant effects of the essential oil from aerial parts of *Mentha* x *villosa* in rats. *Phytomedicine* 2004;11:490–7. doi:10.1016/j.phymed.2004.04.002.

[66] Lahlou S, Carneiro-Leão RFL, Leal-Cardoso JH, Toscano CF. Cardiovascular effects of the essential oil *Mentha* x *villosa* and its main constituent, piperitenone oxide, in normotensive anaesthetised rats: role of the autonomic nervous system. *Planta Med* 2001;67:638–43.

[67] Lahlou S, Magalhães PJC, Carneiro-Leão RFL, Leal-Cardoso JH. Involvement of nitric oxide in the mediation of the hypotensive action of the essential oil of *Mentha* x *villosa* in normotensive conscious rats. *Planta Med* 2002;68:694–9. doi:10.1055/s-2002-33797.

[68] Matos FJDA, Machado MIL, Craveiro AA, Alencar JW, Barbosa JM, da Cunha EVL, et al. Essential oil of *Mentha* x *villosa* Huds. from Northeastern Brazil. *J Essent Oil Res* 1999;11:41–4. doi:10.1080/10412905.1-999.9701066.

[69] Shiva Kumar A, Jeyaprakash K, Chellappan DR, Murugan R. Vasorelaxant and cardiovascular properties of the essential oil of *Pogostemon elsholtzioides*. *J Ethnopharmacol* 2017;199:86–90. doi:10.1016/j.jep.2017.01.036.

[70] Bigliani MC, Rossetti V, Grondona E, Lo Presti S, Paglini PM, Rivero V, et al. Chemical compositions and properties of *Schinus areira* L. essential oil on airway inflammation and cardiovascular system of mice and rabbits. *Food Chem Toxicol* 2012;50:2282–8. doi:10.1016/j.fct.2012.04.025.

[71] Menezes IAC, Barreto CMN, Antoniolli ÂR, Santos MR V, de Sousa DP. Hypotensive activity of terpenes found in essential oils. Zeitschrift Fur Naturforsch - *Sect C J Biosci* 2010;65 C:562–6.

[72] Anjos PJC, Lima AO, Cunha PS, De Sousa DP, Onofre ASC, Ribeiro TP, et al. Cardiovascular effects induced by linalool in normotensive and hypertensive rats. Zeitschrift Für Naturforschung C - *A J Biosci* 2013;68:181–90.

[73] Bastos JFA, Moreira ÍJA, Ribeiro TP, Medeiros IA, Antoniolli AR, De Sousa DP, et al. Hypotensive and vasorelaxant effects of citronellol, a monoterpene alcohol, in rats. *Basic Clin Pharmacol Toxicol* 2010;106:331–7. doi:10.1111/j.1742-7843.2009.00492.x.

[74] de Menezes-Filho JER, Gondim ANS, Cruz JS, de Souza AA, Santos JNA Dos, Conde-Garcia EA, et al. Geraniol blocks calcium and potassium channels in the mammalian myocardium: useful effects to treat arrhythmias. *Basic Clin Pharmacol Toxicol* 2014;115:534–44. doi:10.1111/bcpt.12274.

[75] Guedes DN, Silva DF, Barbosa-Filho JM, Medeiros IA. Muscarinic agonist properties involved in the hypotensive and vasorelaxant responses of rotundifolone in rats. *Planta Med* 2002;68:700–4. doi:10.1055/s-2002-33795.

[76] Lahlou S, Figueiredo AF, Magalhães PJC, Leal-Cardoso JH. Cardiovascular effects of 1,8-cineole, a terpenoid oxide present in many plant essential oils, in normotensive rats. *Can J Physiol Pharmacol* 2002;80:1125–31. doi:10.1139/y02-142.

[77] Ribeiro TP, Porto DL, Menezes CP, Antunes AA, Silva DF, De Souza DP, et al. Unraveling the cardiovascular effects induced by α-terpineol: a role for the NO-cGMP pathway. *Clin Exp Pharmacol Physiol* 2010;37:811–6. doi:10.1111/j.1440-1681.2010.05383.x.

[78] Lahlou S, Interaminense LFL, Leal-Cardoso JH, Duarte GP. Antihypertensive effects of the essential oil of *Alpinia zerumbet* and its main constituent, terpinen-4-ol, in DOCA-salt hypertensive conscious rats. *Fundam Clin Pharmacol* 2003;17:323–30. doi:10.1046/j.1472-8206.2003.00150.x.

[79] Dantas BPV, Alves QL, de Assis KS, Ribeiro TP, de Almeida MM, de Vasconcelos AP, et al. Participation of the TRP channel in the cardiovascular effects induced by carvacrol in normotensive rat.

Vascul Pharmacol 2015;67–69:48–58. doi:10.1016/j.vph. 2015.02.016.

[80] Aydin Y, Kutlay Ö, Ari S, Duman S, Uzuner K, Aydin S. Hypotensive effects of carvacrol on the blood pressure of normotensive rats. *Planta Med* 2007;73:1365–71. doi:10.1055/s-2007-990236.

[81] Krams R, Bäck M. *The ESC textbook of vascular biology*. n.d.

[82] Santos BA, Roman-Campos D, Carvalho MS, Miranda FMF, Carneiro DC, Cavalcante PH, et al. Cardiodepressive effect elicited by the essential oil of *Alpinia speciosa* is related to L-type Ca 2^+ current blockade. *Phytomedicine* 2011;18:539–43. doi:10.1016/ j.phymed.2010.10.015.

[83] Pinto N V, Assreuy AMS, Coelho-de-Souza AN, Ceccatto VM, Magalhães PJC, Lahlou S, et al. Endothelium-dependent vasorelaxant effects of the essential oil from aerial parts of *Alpinia zerumbet* and its main constituent 1,8-cineole in rats. *Phytomedicine* 2009;16:1151–5. doi:10.1016/j.phymed. 2009.04.007.

[84] Tao L, Hu HS, Shen XC. Endothelium-dependent vasodilatation effects of the essential oil from Fructus Alpiniae Zerumbet (EOFAZ) on rat thoracic aortic rings in vitro. *Phytomedicine* 2013;20:387–93. doi:10.1016/j.phymed.2012.12.014.

[85] Dib I, Fauconnier M-L, Sindic M, Belmekki F, Assaidi A, Berrabah M, et al. Chemical composition, vasorelaxant, antioxidant and antiplatelet effects of essential oil of *Artemisia campestris* L. from Oriental Morocco. *BMC Complement Altern Med* 2017;17:82. doi:10.1186/s12906-017-1598-2.

[86] Spadaro F, Costa R, Circosta C, Occhiuto F. Volatile composition and biological activity of key lime *Citrus aurantifolia* essential oil. *Nat Prod Commun* 2012;7:1523–6.

[87] Kang P, Ryu KII, Lee JM, Kim HK, Seol GH. Endothelium- and smooth muscle-dependent vasodilator effects of *Citrus aurantium* L. var. *amara*: Focus on $Ca2^+$ modulation. *Biomed Pharmacother* 2016;82:467–71. doi:10.1016/j.biopha.2016.05.030.

[88] Kang P, Suh SH, Min SS, Seol GH. The essential oil of *Citrus bergamia* Risso induces vasorelaxation of the mouse aorta by activating K(+) channels and inhibiting Ca(2+) influx. *J Pharm Pharmacol* 2013;65:745–9. doi:10.1111/jphp.12031.

[89] de França-Neto A, Cardoso-Teixeira AC, Medeiros TC, Quinto-Farias M do S, Sampaio CM de S, Coelho-de-Souza AN, et al. Essential oil of *Croton argyrophylloides*: toxicological aspects and vasorelaxant activity in rats. *Nat Prod Commun* 2012;7:1397–400.

[90] Lahlou S, Leal-Cardoso JH, Magalhães PJ. Essential oil of *Croton nepetaefolius* decreases blood pressure through an action upon vascular smooth muscle: studies in DOCA-salt hypertensive rats. *Planta Med* 2000;66:138–43. doi:10.1055/s-2000-11133.

[91] Lahlou S, Leal-Cardoso JH, Magalhães PJC, Coelho-de-Souza AN, Duarte GP. Cardiovascular effects of the essential oil of *Croton nepetaefolius* in rats: Role of the Autonomic Nervous System. *Planta Med* 1999;65:553–7. doi:10.1055/s-1999-14025.

[92] Magalhães PJC, Lahlou S, Jucá DM, Coelho-de-Souza LN, da Frota PTT, da Costa AMG, et al. Vasorelaxation induced by the essential oil of *Croton nepetaefolius* and its constituents in rat aorta are partially mediated by the endothelium. *Fundam Clin Pharmacol* 2008;22:169–77. doi:10.1111/j.1472-8206.2008.00571.x.

[93] Martinsen A, Baccelli C, Navarro I, Abad A, Quetin-Leclercq J, Morel N. Vascular activity of a natural diterpene isolated from *Croton zambesicus* and of a structurally similar synthetic trachylobane. *Vascul Pharmacol* 2010;52:63–9. doi:10.1016/j.vph.2009.11.002.

[94] Silva FS, Menezes PMN, de Sá PGS, Oliveira AL de S, Souza EAA, Almeida JRG da S, et al. Chemical composition and pharmacological properties of the essential oils obtained seasonally from *Lippia thymoides*. *Pharm Biol* 2016;54:25–34. doi:10.3109/13880209.2015.1005751.

[95] Pereira SL, Marques AM, Sudo RT, Kaplan MAC, Zapata-Sudo G. Vasodilator activity of the essential oil from aerial parts of *Pectis*

brevipedunculata and its main constituent citral in rat aorta. *Molecules* 2013;18:3072–85. doi:10.3390/molecules18033072.

[96] Zadeh GS, Panahi N. Endothelium-independent vasorelaxant activity of *Trachyspermum ammi* essential oil on rat aorta. *Clin Exp Hypertens* 2017;39:133–8. doi:10.1080/10641963.2016.1235178.

[97] de C. Correia AC, Ferreira TF, Martins IRR, Macêdo CL, de S. Monteiro F, Costa VCO, et al. Essential oil from the leaves of *Xylopia langsdorfiana* (Annonaceae) as a possible spasmolytic agent. *Nat Prod Res* 2015;29:980–4. doi:10.1080/14786419.2014.964706.

[98] Peixoto-Neves D, Silva-Alves KS, Gomes MDM, Lima FC, Lahlou S, Magalhães PJC, et al. Vasorelaxant effects of the monoterpenic phenol isomers, carvacrol and thymol, on rat isolated aorta. *Fundam Clin Pharmacol* 2010;24:341–50. doi:10.1111/j.1472-8206.2009.00768.x.

[99] Shabir H, Kundu S, Basir SF, Khan LA. Modulation of Pb(II) caused aortal constriction by eugenol and carvacrol. *Biol Trace Elem Res* 2014;161:116–22. doi:10.1007/s12011-014-0081-x.

[100] Baccelli C, Martinsen A, Morel N, Quetin-Leclercq J. Vasorelaxant activity of essential oils from *Croton zambesicus* and some of their constituents. *Planta Med* 2010;76:1506–11. doi:10.1055/s-0030-1249820.

[101] Kundu S, Shabir H, Basir SF, Khan LA. Inhibition of As(III) and Hg(II) caused aortic hypercontraction by eugenol, linalool and carvone. *J Smooth Muscle Res* 2014;50:93–102. doi:10.1540/jsmr.50.93.

[102] Soares MCMS, Damiani CEN, Moreira CM, Stefanon I, Vassallo D V. Eucalyptol, an essential oil, reduces contractile activity in rat cardiac muscle. Brazilian *J Med Biol Res = Rev Bras Pesqui Medicas E Biol* 2005;38:453–61. doi:/S0100-879X2005000300017.

[103] Guedes DN, Silva DF, Barbosa-Filho JM, Medeiros IA. Calcium antagonism and the vasorelaxation of the rat aorta induced by rotundifolone. *Brazilian J Med Biol Res* 2004;37:1881–7. doi:/S0100-879X2004001200014.

[104] Maia-Joca RPM, Joca HC, Ribeiro FJP, Nascimento RV Do, Silva-Alves KS, Cruz JS, et al. Investigation of terpinen-4-ol effects on vascular smooth muscle relaxation. *Life Sci* 2014;115:52–8. doi:10.1016/j.lfs.2014.08.022.

[105] Koto R, Imamura M, Watanabe C, Obayashi S, Shiraishi M, Sasaki Y, et al. Linalyl acetate as a major ingredient of lavender essential oil relaxes the rabbit vascular smooth muscle through dephosphorylation of myosin light chain. *J Cardiovasc Pharmacol* 2006;48:850–6. doi:10.1097/01.fjc.0000238589.00365.42.

[106] Earley S, Gonzales AL, Garcia ZI. A dietary agonist of transient receptor potential cation channel V3 elicits endothelium-dependent vasodilation. *Mol Pharmacol* 2010;77:612–20. doi:10.1124/mol.109.060715.

[107] Cheang WS, Lam MY, Wong WT, Tian XY, Lau CW, Zhu Z, et al. Menthol relaxes rat aortae, mesenteric and coronary arteries by inhibiting calcium influx. *Eur J Pharmacol* 2013;702:79–84. doi:10.1016/j.ejphar.2013.01.028.

[108] Johnson CD, Melanaphy D, Purse A, Stokesberry SA, Dickson P, Zholos A V. Transient receptor potential melastatin 8 channel involvement in the regulation of vascular tone. *Am J Physiol - Hear Circ Physiol* 2009;296:H1868–77. doi:10.1152/ajpheart. 01112. 2008.

[109] de Siqueira RJB, Freire WBS, Vasconcelos-Silva A a., Fonseca-Magalhães P a., Lima FJB, Brito TS, et al. In vitro characterization of the pharmacological effects induced by (–)-α-bisabolol in rat smooth muscle preparations. *Can J Physiol Pharmacol* 2012;90:23–35. doi:10.1139/y11-094.

[110] de Siqueira RJB, Ribeiro-Filho H V., Freire RS, Cosker F, Freire WBS, Vasconcelos-Silva AA, et al. (-)-α-Bisabolol inhibits preferentially electromechanical coupling on rat isolated arteries. *Vascul Pharmacol* 2014;63:37–45. doi:10.1016/j.vph.2014.06.006.

[111] El Tantawy ME, El Sakhawy FS, El Sohly MA, Ross SA. Chemical Composition and biological activity of the essential oil of the fruit

of *Taxodium distichum* L. Rich growing in Egypt. *J Essent Oil Res* 1999;11:386–92. doi:10.1080/10412905.1999.9701161.

[112] Reid IR, Birstow SM, Bolland MJ. Calcium and cardiovascular disease. *Endocrinol Metab* (Seoul, Korea) 2017;32:339–49. doi:10.3803/EnM.2017.32.3.339.

[113] Marks AR. Calcium and the heart: a question of life and death. *J Clin Invest* 2003;111:597–600. doi:10.1172/JCI18067.

[114] You JH, Kang P, Min SS, Seol GH. Bergamot essential oil differentially modulates intracellular $Ca2^{+}$ levels in vascular endothelial and smooth muscle cells. *J Cardiovasc Pharmacol* 2013;61:324–8. doi:10.1097/FJC.0b013e3182834681.

[115] Maiwulanjiang M, Bi CWC, Lee PSC, Xin G, Miernisha A, Lau KM, et al. The volatile oil of N*ardostachyos* radix et rhizoma induces endothelial nitric oxide synthase cctivity in HUVEC cells. *PLoS One* 2015;10:e0116761. doi:10.1371/journal.pone.0116761.

[116] Magyar J, Szentandrássy N, Bányász T, Fülöp L, Varró A, Nánási PP. Effects of terpenoid phenol derivatives on calcium current in canine and human ventricular cardiomyocytes. *Eur J Pharmacol* 2004;487:29–36. doi:10.1016/j.ejphar.2004.01.011.

[117] Magyar J, Szentandrássy N, Bányász T, Fülöp L, Varró A, Nánási PP. Effects of thymol on calcium and potassium currents in canine and human ventricular cardiomyocytes. *Br J Pharmacol* 2002;136:330–8. doi:10.1038/sj.bjp.0704718.

[118] Szentandrássy N, Szigeti G, Szegedi C, Sárközi S, Magyar J, Bányász T, et al. Effect of thymol on calcium handling in mammalian ventricular myocardium. *Life Sci* 2004;74:909–21. doi:10.1016/j.lfs.2003.09.034.

[119] Sensch O, Vierling W, Brandt W, Reiter M. Effects of inhibition of calcium and potassium currents in guinea-pig cardiac contraction: comparison of beta-caryophyllene oxide, eugenol, and nifedipine. *Br J Pharmacol* 2000;131:1089–96. doi:10.1038/sj.bjp.0703673.

[120] Nascimento NRF, Leal-Cardoso JH, Lessa LMA, Roriz-Filho JS, Cunha KMA, Fonteles MC. Terpinen-4-ol: mechanisms of

relaxation on rabbit duodenum. *J Pharm Pharmacol* 2005;57:467–74. doi:10.1211/0022357055696.

[121] Astudillo A, Hong E, Bye R, Navarrete A. Antispasmodic activity of extracts and compounds of *Acalypha phleoides* Cav. *Phytother Res* 2004;18:102–6. doi:10.1002/ptr.1414.

[122] Willoughby S, Holmes A, Loscalzo J. Platelets and cardiovascular disease. *Eur J Cardiovasc Nurs* 2002;1:273–88.

[123] Sangkuhl K, Shuldiner AR, Klein TE, Altman RB. Platelet aggregation pathway. *Pharmacogenet Genomics* 2011;21:516–21. doi:10.1097/FPC.0b013e3283406323.

[124] Tognolini M, Barocelli E, Ballabeni V, Bruni R, Bianchi A, Chiavarini M, et al. Comparative screening of plant essential oils: phenylpropanoid moiety as basic core for antiplatelet activity. *Life Sci* 2006;78:1419–32. doi:10.1016/j.lfs.2005.07.020.

[125] Yan Y, O W, Zhao X, Ye X, Zhang C, Hao J, et al. Effect of essential oil of *Syringa pinnatifolia* Hemsl. var. *alashanensis* on ischemia of myocardium, hypoxia and platelet aggregation. *J Ethnopharmacol* 2010;131:248–55. doi:10.1016/j.jep.2010.06.027.

[126] Reiner Z, Catapano AL, De Backer G, Graham I, Taskinen M-R, Wiklund O, et al. ESC/EAS Guidelines for the management of dyslipidaemias: The Task Force for the management of dyslipidaemias of the European Society of Cardiology (ESC) and the European Atherosclerosis Society (EAS). *Eur Heart J* 2011;32:1769–818. doi:10.1093/eurheartj/ehr158.

[127] Racette SB, Lin X, Lefevre M, Spearie CA, Most MM, Ma L, et al. Dose effects of dietary phytosterols on cholesterol metabolism: a controlled feeding study. *Am J Clin Nutr* 2010;91:32–8. doi:10.3945/ajcn.2009.28070.

[128] Charlton-Menys V, Durrington PN. Human cholesterol metabolism and therapeutic molecules. *Exp Physiol* 2008;93:27–42. doi:10.1113/expphysiol.2006.035147.

[129] Morgan AE, Mooney KM, Wilkinson SJ, Pickles NA, Mc Auley MT. Cholesterol metabolism: A review of how ageing disrupts the

biological mechanisms responsible for its regulation. *Ageing Res Rev* 2016;27:108–24. doi:10.1016/J.ARR.2016.03.008.

[130] Singh V, Jain M, Misra A, Khanna V, Rana M, Prakash P, et al. Curcuma oil ameliorates hyperlipidaemia and associated deleterious effects in golden Syrian hamsters. *Br J Nutr* 2013;110:437–46. doi:10.1017/S0007114512005363.

[131] Kim J-H, Lee H-J, Jeong S-J, Lee M-H, Kim S-H. Essential oil of *Pinus koraiensis* leaves exerts antihyperlipidemic effects via up-regulation of low-density lipoprotein receptor and inhibition of acyl-coenzyme A: cholesterol acyltransferase. *Phytother Res* 2012;26:1314–9. doi:10.1002/ptr.3734.

[132] Chung MJ, Park KW, Kim KH, Kim C-T, Baek JP, Bang K-H, et al. Asian plantain (*Plantago asiatica*) essential oils suppress 3-hydroxy-3-methyl-glutaryl-co-enzyme A reductase expression in vitro and in vivo and show hypocholesterolaemic properties in mice. *Br J Nutr* 2008;99:67–75. doi:10.1017/S0007114507798926.

[133] Saravanan S, Pari L. Role of thymol on hyperglycemia and hyperlipidemia in high fat diet-induced type 2 diabetic C57BL/6J mice. *Eur J Pharmacol* 2015;761:279–87. doi:10.1016/j.ejphar.2015.05.034.

[134] Ezhumalai M, Ashokkumar N, Pugalendi KV. Combination of carvacrol and rosiglitazone ameliorates high fat diet induced changes in lipids and inflammatory markers in C57BL/6J mice. *Biochimie* 2015;110:129–36. doi:10.1016/j.biochi.2014.12.005.

[135] Galle M, Kladniew BR, Castro MA, Villegas SM, Lacunza E, Polo M, et al. Modulation by geraniol of gene expression involved in lipid metabolism leading to a reduction of serum-cholesterol and triglyceride levels. *Phytomedicine* 2015;22:696–704. doi:10.1016/j.phymed.2015.04.005.

[136] Jayachandran M, Chandrasekaran B, Namasivayam N. Effect of geraniol, a plant derived monoterpene on lipids and lipid metabolizing enzymes in experimental hyperlipidemic hamsters. *Mol Cell Biochem* 2015;398:39–53. doi:10.1007/s11010-014-2203-3.

[137] Vallianou I, Peroulis N, Pantazis P, Hadzopoulou-Cladaras M. Camphene, a plant-derived monoterpene, reduces plasma cholesterol and triglycerides in hyperlipidemic rats independently of HMG-CoA reductase activity. *PLoS One* 2011;6:e20516. doi:10.1371/journal.pone.0020516.

[138] Naderi GA, Asgary S, Ani M, Sarraf-Zadegan N, Safari MR. Effect of some volatile oils on the affinity of intact and oxidized low-density lipoproteins for adrenal cell surface receptors. *Mol Cell Biochem* 2004;267:59–66.

[139] Ibanez B, James S, Agewall S, Antunes MJ, Bucciarelli-Ducci C, Bueno H, et al. 2017 ESC Guidelines for the management of acute myocardial infarction in patients presenting with ST-segment elevation. *Eur Heart J* 2017;39:119–77. doi:10.1093/eurheartj/ehx393.

[140] Szűcs G, Murlasits Z, Török S, Kocsis GF, Pálóczi J, Görbe A, et al. Cardioprotection by farnesol: role of the mevalonate pathway. *Cardiovasc Drugs Ther* 2013;27:269–77. doi:10.1007/s10557-013-6460-2.

[141] Shen X-C, Tao L, Li W-K, Zhang Y-Y, Luo H, Xia Y-Y. Evidence-based antioxidant activity of the essential oil from Fructus A. zerumbet on cultured human umbilical vein endothelial cells' injury induced by ox-LDL. *BMC Complement Altern Med* 2012;12:174. doi:10.1186/1472-6882-12-174.

[142] Xiao T, Zeng Y, Xu Y, Zhang Y, Jiang Y, Tao L, et al. The endothelial protective properties of essential oil from Fructus Alpiniae zerumbet via the Akt/NOS-NO signaling pathway in vitro. *Planta Med* 2014;80:1628–34. doi:10.1055/s-0034-1383129.

[143] Danesi F, Elementi S, Neri R, Maranesi M, D'Antuono LF, Bordoni A. Effect of cultivar on the protection of cardiomyocytes from oxidative stress by essential oils and aqueous extracts of basil (*Ocimum basilicum* L.). *J Agric Food Chem* 2008;56:9911–7. doi:10.1021/jf8018547.

[144] Lee KP, Sudjarwo GW, Jung SH, Lee D, Lee DY, Lee GB, et al. Carvacrol inhibits atherosclerotic neointima formation by

downregulating reactive oxygen species production in vascular smooth muscle cells. *Atherosclerosis* 2015;240:367–73. doi:10.1016/j.atherosclerosis.2015.03.038.

[145] Höferl M, Krist S, Buchbauer G. Chirality influences the effects of linalool on physiological parameters of stress. *Planta Med* 2006;72:1188–92. doi:10.1055/s-2006-947202.

[146] Liu F, Huang Z-Z, Sun Y-H, Li T, Yang D-H, Xu G, et al. Four main active ingredients derived from a traditional chinese medicine guanxin shutong capsule cause cardioprotection during myocardial ischemia injury calcium overload suppression. *Phytother Res* 2017;31:507–15. doi:10.1002/ptr.5787.

[147] Baddar NWA-H, Aburjai T a, Taha MO, Disi AM. Thujone corrects cholesterol and triglyceride profiles in diabetic rat model. *Nat Prod Res* 2011;25:1180–4. doi:10.1080/14786419.2010.496116.

[148] Basha RH, Sankaranarayanan C. β-Caryophyllene, a natural sesquiterpene lactone attenuates hyperglycemia mediated oxidative and inflammatory stress in experimental diabetic rats. *Chem Biol Interact* 2016;245:50–8. doi:10.1016/j.cbi.2015.12.019.

[149] Chompoo J, Upadhyay A, Gima S, Fukuta M, Tawata S. Antiatherogenic properties of acetone extract of *Alpinia zerumbet* seeds. *Molecules* 2012;17:6237–48. doi:10.3390/molecules17066237.

[150] Chuang C-M, Wang H-E, Peng C-C, Chen KC, Peng RY. Hypolipidemic effects of different angiocarp parts of *Alpinia zerumbet*. *Pharm Biol* 2011;49:1257–64. doi:10.3109/13880209.2011.589856.

[151] Garcia D, Domingues MV, Rodrigues E. Ethnopharmacological survey among migrants living in the southeast atlantic forest of Diadema, São Paulo, Brazil. *J Ethnobiol Ethnomed* 2010;6:29. doi:10.1186/1746-4269-6-29.

[152] Wubetu M, Abula T, Dejenu G. Ethnopharmacologic survey of medicinal plants used to treat human diseases by traditional medical practitioners in Dega Damot district, Amhara, Northwestern

Ethiopia. *BMC Res Notes* 2017;10:157. doi:10.1186/s13104-017-2482-3.

[153] Cruz EC, Andrade-Cetto A. Ethnopharmacological field study of the plants used to treat type 2 diabetes among the Cakchiquels in Guatemala. *J Ethnopharmacol* 2015;159:238–44. doi:10.1016/J.JEP.2014.11.021.

[154] Maciel MAM, Pinto AC, Arruda AC, Pamplona SGS., Vanderlinde FA, Lapa AJ, et al. Ethnopharmacology, phytochemistry and pharmacology: a successful combination in the study of *Croton cajucara*. *J Ethnopharmacol* 2000;70:41–55. doi:10.1016/S0378-8741(99)00159-2.

[155] Lekshmi PC, Arimboor R, Indulekha PS, Nirmala Menon A. Turmeric (*Curcuma longa* L.) volatile oil inhibits key enzymes linked to type 2 diabetes. *Int J Food Sci Nutr* 2012;63:832–4. doi:10.3109/09637486.2011.607156.

[156] Adaramoye OA, Anjos RM, Almeida MM, Veras RC, Silvia DF, Oliveira FA, et al. Hypotensive and endothelium-independent vasorelaxant effects of methanolic extract from *Curcuma longa* L. in rats. *J Ethnopharmacol* 2009;124:457–62. doi:10.1016/J.JEP.2009.05.021.

INDEX

B

C

D

G

H

I

N

O

P

R

S

T

V

W

X

Y

Z